EXPLORING *the* DYNAMIC EARTH

Michelle K. Hall
Science Education Solutions, Inc.

C. Scott Walker
Oregon State University

Anne K. Huth
Science Education Solutions, Inc.

Robert F. Butler
The University of Portland

Larry P. Kendall
Science Education Solutions, Inc.

Jeff S. Jenness
Jenness Enterprises

THOMSON
BROOKS/COLE™

Australia • Brazil • Canada • Mexico • Singapore
Spain • United Kingdom • United States

THOMSON
BROOKS/COLE

Exploring the Dynamic Earth

Michelle K. Hall
C. Scott Walker
Anne K. Huth
Robert F. Butler
Larry P. Kendall
Jeff S. Jenness

Executive Editor: Peter Adams
Assistant Editor: Carol Benedict
Editorial Assistant: Anna Jarzab
Marketing Manager: Joe Rogove
Marketing Communications Manager: Bryan Vann
Content Project Manager: Belinda Krohmer
Creative Director: Rob Hugel
Art Director: Vernon Boes
Print Buyer: Nora Massuda
Permissions Editor: Roberta Broyer
Cover Designer: Denise Davidson
Cover Image: Digital Vision
Cover Printer: Courier-Kendallville
Printer: Courier-Kendallville

Printed in the United States of America
1 2 3 4 5 6 7 10 09 08 07 06

Library of Congress Control Number: 2006906785

ISBN 0495115096

Thomson Higher Education
10 Davis Drive
Belmont, CA 94002-3098
USA

Acknowledgments

Series Authors

Michelle K. Hall
Anne K. Huth
C. Scott Walker
Jennifer A. Weeks
Robert F. Butler
Jeff S. Jenness
Larry P. Kendall

Programmers

Jeff S. Jenness
Christian J. Schaller

Media Developers

Christian J. Schaller
Ted Stude

Technical Reviewers

Margaret Chernosky
Lydia Fox
Kathleen Friedman
Robert Kolvoord
Carla McAuliffe
Anne Ortiz
Alan Sills
Jessica Walker

Scientific Reviewers

Robert Butler
Peter Kresan
David Smith
Terry Wallace
Joseph Watkins

Technical Advisors

Tom Garrison
Cheryl Greengrove
Miles Logsdon
David Smith
Jim Washburne
Douglas Yarger

Proofreader

James Kevin

Data Providers

Robert Diaz
Virginia Institute of Marine Science
Don Pool
USGS Tucson Field Office
Pima County Department of Transportation Technical Services Division
Nancy Rabelais
Louisiana University Marine Consortium
Christopher Scotese
The Paleomap Project

Field Testers

Tekla Cook
Jo Dodds
Christine Donovan
Lillian Fox
Joshua Hall
Alan Kelley
Bob Kolvoord
Kathy Krucker
Kathy Likos
Daniel Montoya
Stephen Murray
Anthony Ochiuzzi
Graciela Rendon-Coke
Richard Spitzer
Anne Thames
Steven Uyeda
Margaret Wilch

Assistants

Tammy Baldwin
Christine Hallman
Sara McNamara
Marie Renwald
Megan Sayles

The authors would also like to thank the many scientists who took the time to learn about this project and share critical research data and expertise, and to the agencies and individuals that have given us permission to include their outstanding illustrations and photos.

SAGUARO
Science And GIS Unlocking Analysis & Research Opportunities
The SAGUARO Project

Michelle K. Hall, Director
Science Education Solutions, Inc.
Los Alamos Research Park
4200 West Jemez Drive
Synergy Center Suite 301
Los Alamos NM 87544-2587

saguaro@scieds.com

Science Education Solutions
understanding our world through science
http://www.scieds.com/saguaro

Table of contents

Introduction

Investigations

Getting started

Software needed

The investigations in this version of the GIS Investigations for the Earth Sciences series are intended for use on campuses with ArcGIS 9.x laboratory or site licenses. All necessary software and data files should be set up and ready for you to use on campus or lab computers. Your instructor will provide you with information on the location(s) of these computers, and where to access the data files for the investigations.

If you have a valid license for the student edition of ArcGIS 9.x, you may be able obtain the necessary data, additional software, and installation instructions to install and use these materials on your personal computer. Contact your instructor for information on where to obtain the required files.

In addition to ArcGIS 9.x, these materials require the following software. Most are available as free downloads from their respective publishers.

- Saguaro Tools for ArcGIS 9.x (contact your instructor)
- Web browser (Internet Explorer, Netscape, Opera, Mozilla, etc.)
- Windows Media Player
- Google Earth
- Adobe Reader

What you need to know

The authors of this book assume that you have the following computer skills:

- Turning the computer on and, if necessary, logging in as a user.
- Navigating the file system to find folders, applications, and files.
- Launching applications and opening files.
- Opening, closing, moving, and resizing windows.
- Using tools, buttons, menus, and dialog boxes.

Other than these basic computer skills, students and instructors do not need prior experience using ArcGIS software. We do, however, strongly recommend that instructors run through each investigation before presenting it to students, to familiarize themselves with the data and techniques. We also strongly recommend that students and instructors:

- Be familiar with visual cues used in these materials.
- Read all instructions carefully.
- Pay attention to information provided in computer-screen shots and in the left-hand sidebar on each page.

Attention course instructors!

The Instructor Guide (answer key), investigation data, additional software, installation instructions, troubleshooting guide, and other resources can be downloaded from the Instructor Resource Center at:

http://www.thomsonedu.com/earthscience/hall/earth

You will need a user name and password, provided by the publisher, to access these files.

Mac OS X compatibility

At this time, ArcGIS 9.x is not compatible with any version of the Macintosh operating system. Users may get satisfactory, but slow, performance running ArcGIS under Virtual PC software on a Macintosh computer, but this is neither recommended nor supported.

What is the difference between ArcGIS, ArcView, and ArcMap?

For all practical purposes, you can think of these as three different names for the same geographic information system software.

Technically, ArcGIS is a family of related tools for managing geographic information systems on a variety of scales, ranging from Web servers to handheld devices.

One of these tools is a package for use on desktop computers, called ArcView. The ArcView 9.x software package consists of three components, each interacting with geographic data in a unique way:

- ArcCatalog — management
- ArcMap — viewing and analysis
- ArcToolbox — modifying data

In these materials, you will only be using the ArcMap component of ArcGIS 9.x.

Using ArcMap

Launching ArcMap and opening project files

- To launch the ArcMap application, click the Start button on the Windows Taskbar and choose **All Programs** > **ArcGIS** > **ArcMap** (ArcMap).
- If you see the ArcMap dialog box, choose **Browse for file** under the **An existing map** option.

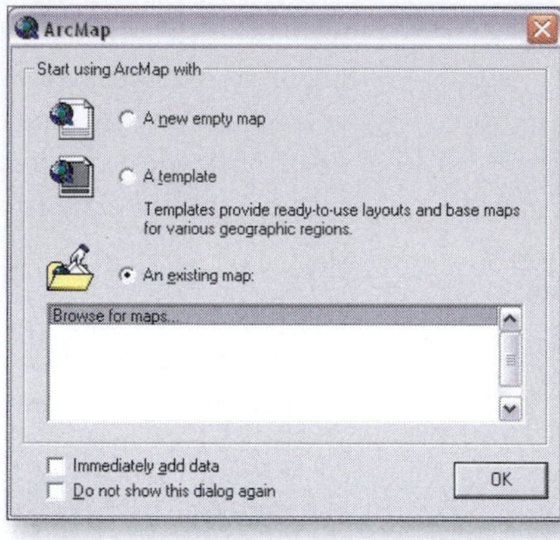

- Choose **File** > **Open**.
- Navigate to the appropriate unit data folder installed on your local hard drive or server and open it. (Your instructor or lab administrator can tell you where to find the investigation folders.)
- Locate the specified ArcMap document file and open it. (The .mxd file extension may or may not be visible, depending on how the computer has been set up.)

ArcMap shortcuts

In addition to the Start menu, an ArcMap shortcut may have been placed on the desktop or taskbar for your convenience.

Why doesn't the ArcMap startup dialog box appear?

If someone has checked the **Do not show this dialog again** checkbox, it will no longer appear each time you launch ArcMap.

Visual cues

Visual cues are used to make the investigation directions easier to follow.

- Text preceded by a computer symbol is an instruction—something to do on the computer.
- Names of tools or buttons are capitalized and are followed by a picture of that item as it appears on screen—for example, the Identify tool.
- The > symbol between boldface words or phrases in text indicates a menu choice. Thus, **File > Open…** means "pull down the File menu and choose Open…"

"File > Open…" means…

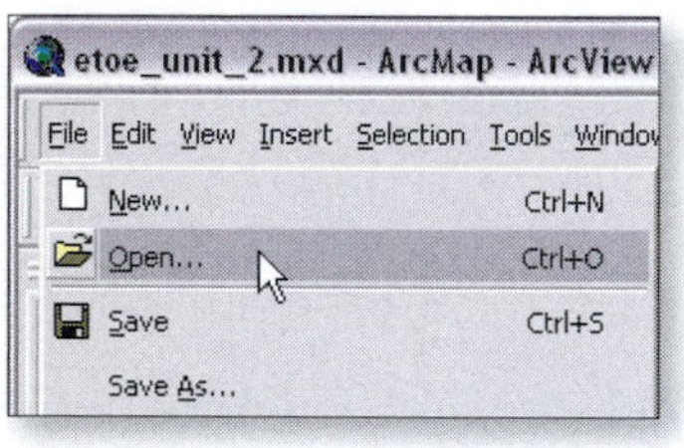

Sidebars contain important information!

The page sidebars contain useful information such as definitions, explanations, illustrations, examples, reminders, warnings, tips, and hints. If you are not sure what to do, look for help in the sidebar first.

Closing map files

When you have completed an investigation or must stop for some reason, choose **File** > **Exit** and click **No** when asked if you want to save your changes.

What if I click "Yes"?

Don't worry if you click **Yes** when you close a map file. The files have been locked to prevent you from accidentally modifying them.

The ArcMap User Interface

Title Bar
Shows the name of the current map file.

SAGUARO Tools
Tools provided by the SAGUARO Project for these materials. Note: This toolbar may appear in a different location.

Menu Bar
Provides menus for performing various operations.

Toolbars
Tools for manipulating and analyzing map data.

Table of Contents
A list of data frames and map layers, and controls for changing the map view.

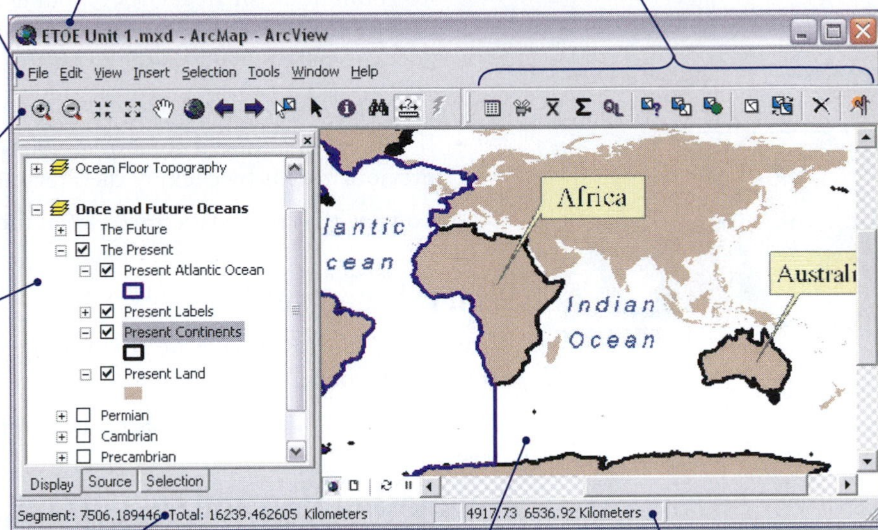

Status Bar
Displays tool descriptions, measurement results, and other information about the current operation.

Data Frame
The currently-selected map.

Coordinates
Displays the coordinates, in the currently-selected units, of the cursor position.

Basic display operations

Activating a data frame

To activate a data frame, right-click its name and choose Activate from the pop-up menu. The title of the activated data frame is highlighted bold.

Expanding and collapsing data frames

To expand a data frame and show its layers, click the expand box ⊞. To collapse a data frame and hide its layers, click the collapse box ⊟.

Selecting layers or layer groups

To select a layer or layer group, click the layer or layer group name. Selected layers or layer groups are highlighted. To select multiple layers or layer groups, hold down the control key while clicking additional names.

Expanding and collapsing layers or layer groups

To expand a layer or layer group, click the expand box ⊞. To collapse a layer or layer group, click the collapse box ⊟.

Turning layers or layer groups on and off

To turn a layer or layer group on, check ☑ the box in front of its name. To turn a layer or layer group off, uncheck the box ☐ in front of its name. If a layer is turned on but is not visible, it may be hidden behind another layer. Try turning off the layers *above* that layer in the Table of Contents.

Table of Contents

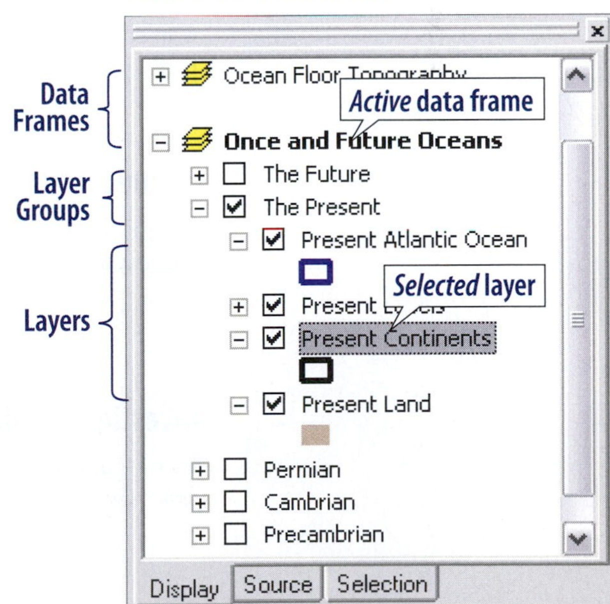

Zooming

The most efficient way to zoom in on a specific area is to drag a box around the area using the Zoom In tool. Drag diagonally from one corner of the area to the opposite corner, then release the button.

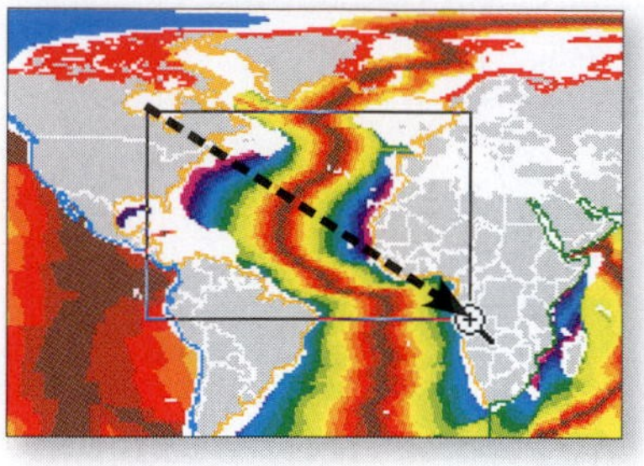

Zooming

ArcMap has tools for zooming—enlarging and reducing areas of the map—that work like the tools you have used in other applications.

- To zoom in on an area, click and drag diagonally with the Zoom In tool to outline the area on the map. When you release the button, the area you selected will rescale to fill the data frame window.
- To zoom out, click anywhere on the map with the Zoom Out tool.
- If you zoom in or out so far that you do not know where you are, undo previous zooms by clicking the Previous Extent button.
- To view the entire data frame, click the Full Extent button.

Want to know more?

If you would like to know more about using ArcGIS, download the **Guide to ArcGIS.pdf** from:

http://www.thomsonedu.com/earthscience/hall/earth

Troubleshooting and support options

ArcGIS help

This module provides all of the directions you need to complete the investigations using ArcGIS 9.x. If you have other questions about the capabilities of ArcGIS, choose **Help > ArcGIS Desktop Help.**

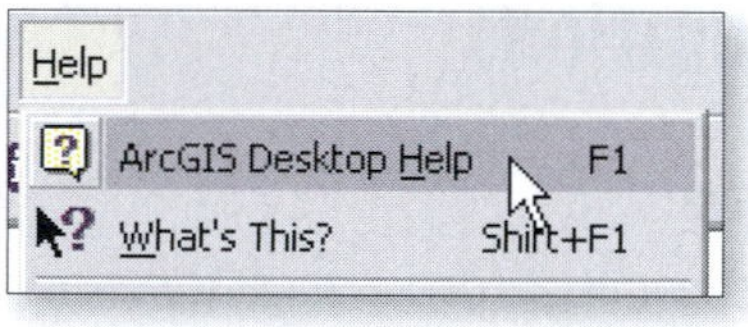

Troubleshooting

The current version of the troubleshooting guide for this series (**Troubleshooting.pdf**) is available from the Instructor Resource Center at:

http://www.thomsonedu.com/earthscience/hall/earth

Technical support

If you experience problems with these materials, contact the Thomson Learning technical support team at:

tl.support@thomson.com

1-800-4230563

Investigation data, updates, and resources

Investigation data, updates, and additional resources related to the *GIS Investigations for the Earth Sciences* series can be downloaded from:

http://www.thomsonedu.com/earthscience/hall/earth

Working with large numbers

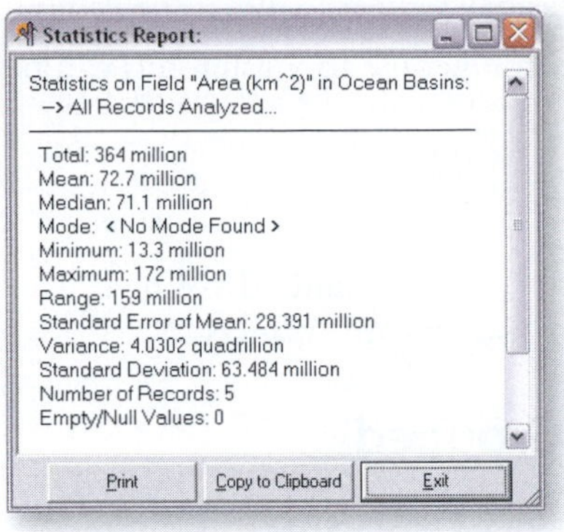

Some of the numbers you will work with in these investigations are quite large. When talking about the amount of water in the ocean or the energy of an earthquake or hurricane, you routinely use values in the billions or even trillions. Where possible, ArcGIS has been modified to make these very large and very small numbers easier to read. For example, in the Statistics Report window shown at left, the total area is given as 364 million, rather than 363958342077361 square kilometers.

Occasionally, you will need to convert millions to billions or thousands, or vice versa. For example, to convert the **Mean** value in the window at left from millions to billions, move the decimal point three places to the *left*. To go from millions to thousands, move the decimal three places to the *right*.

72700 thousand = **72.7 million** = 0.0727 billion

Rounding

Most of these numbers are approximations, so it does not make sense to be overly precise when you are calculating or recording them. Look at the number written below, and the place value of each of the digits. Face it—when you are talking about nearly 149 billion of something, who cares about hundred-thousandths, or even tens of millions?

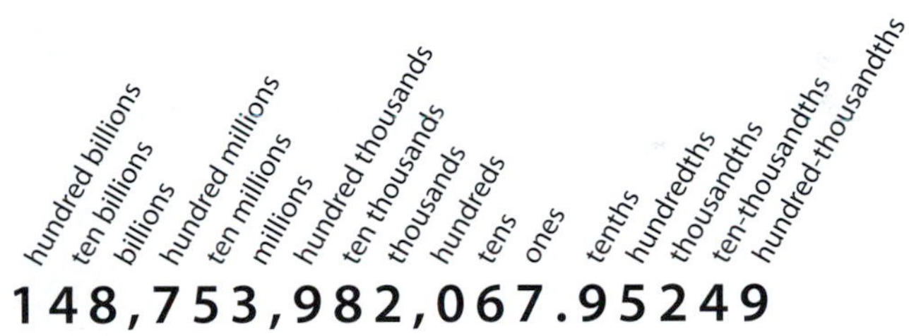

Throughout these investigations, you will be asked to round answers to a particular value and number of decimal places, such as "Round your answer to the nearest 0.1 million." Rounding numbers is simple, if you follow these steps. Examples are shown at the left.

- Look only at the numeral to the right of the place value you are rounding to. For example, when rounding to the nearest thousand, look only at the numeral in the hundreds place.
- If the numeral to the right is 0-4, do not change the number you are rounding to. If the number to the right is 5-9, add one to the number you are rounding to.
- Change whole numerals to the right of the place you are rounding to into zeros, and omit all unused decimal places.
- For any number less than 1, include a zero to the left of the decimal point. (Instead of *.79 billion*, write *0.79 billion.*)

Rounding examples

For example, if your number is

319,740,562.85

To round to the nearest ten million:

- Find the ten millions digit (1).
- Look at the number to its right (9). Because it is between 5 and 9, add one to the ten millions digit to make it 2.
- Change the whole numbers to the right of the ten millions digit to zeros and drop the decimal point and everything to its right. The result is **320,000,000**.

Rounding to the nearest...

...million (1,000,000) = **320,000,000** (adding 1 to 319 gives 320)

...hundred thousand (100,000) = **319,700,000**

...ten thousand (10,000) = **319,740,000**

...thousand (1,000) = **319,741,000**

...hundred (100) = **319,740,600**

...ten (10) = **319,740,560**

...one (1) = **319,740,563**

...tenth (0.1) = **319,740,562.9**

To round to the nearest 0.1 million:

- Find the 0.1 millions digit (7). This is also called the hundred thousands digit.
- Look at the number to its right (4). Because it is between 0 and 4, do not add one to the 0.1 millions digit.
- Insert the decimal point in the proper location. The result is **319.7 million**.

Rounding decimal fractions

Rounding decimals works the same way, except that you are rounding to tenths, hundredths, thousandths, and so on. Do not add zeros to the right of the decimal point. In other words, rounding 2.587 to the nearest tenth is 2.6, *not* 2.600.

Estimating percent area

You will occasionally be asked to estimate the percent area covered by land, ocean, or some other feature. This is a difficult skill for some people to master, but can be learned with practice.

Comparing to standards

One method of estimating coverage is to compare to visual standards. When estimating coverage you need to consider how the features are arranged.

Cloud cover exercise

Here is a simple activity that demonstrates the confusing nature of cover estimates.

- Take two full sheets of blue paper and one of white paper. The blue paper represents sky, and the white paper represents clouds.
- Cut the white sheet in half. Tear or cut the first half of the white sheet into large pieces and glue them onto one of the blue sheets without overlapping.
- Repeat the step above with the other half of the white sheet and the other blue sheet. This time, cut or tear the white sheet into small chunks before gluing them on.

In both cases, the cloud cover is 50 percent. Half of the blue sky is covered by white clouds, but the sheet covered by large clouds appears more open than the sheet covered by small clouds.

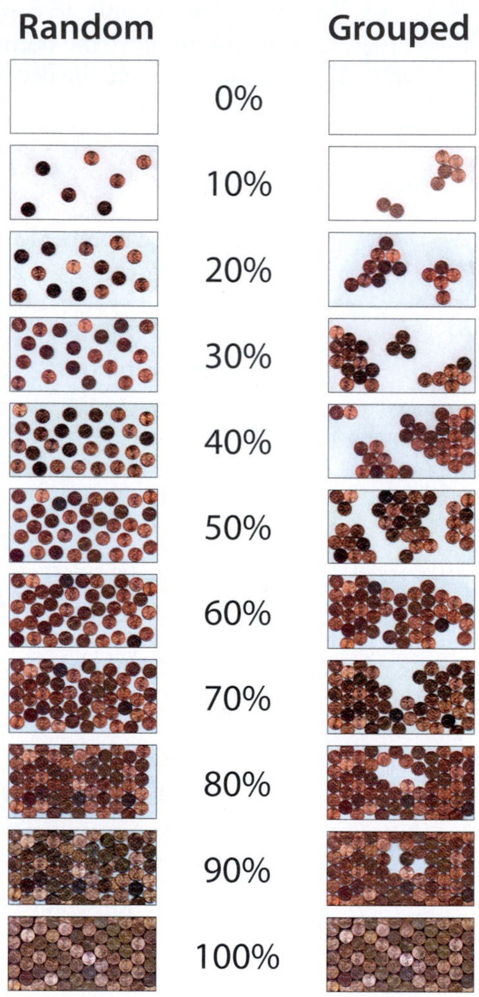

Gridding

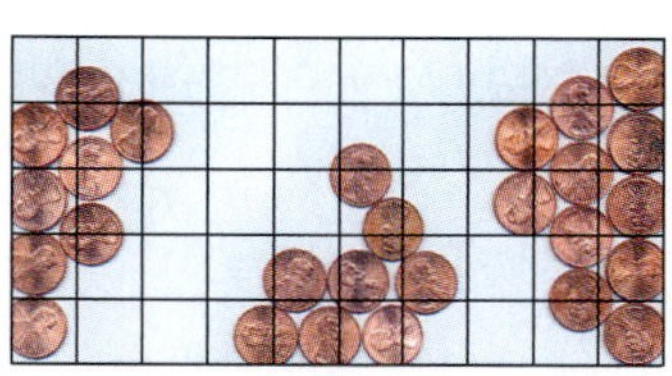

Another approach to estimating coverage is to divide the area up into a grid, either mentally or physically, and determine the number of grid squares that are at least half-covered. To find the percent coverage, calculate the ratio of covered squares to total squares and multiply by 100.

In the example at left, approximately 20 of the 50 squares are at least half covered.

20/50 × 100 = 40% coverage

Philosophy & design

Philosophy

Thinking scientifically

An Earth scientist makes a living by observing and measuring nature. Whether recording and analyzing earthquakes or measuring subtle changes in sea surface temperature over many decades, a successful Earth scientist relies heavily on his or her ability to recognize patterns. Patterns in space and time are the keys to many of the great discoveries about how Earth works. The investigations in this series are designed to help you develop your ability to recognize and interpret nature's fundamental patterns by exploring recent scientific data using a computer and geographic information system (GIS) software.

What is a GIS?

GIS provides tools for organizing, manipulating, analyzing, and visualizing information about the world using digital maps and databases.

Most of these patterns are presented through maps, which are among scientists' most important tools. Maps allow you to visually explore spatial relationships between phenomena such as surface winds and ocean currents; natural features such as continents and ocean basins; and human features such as countries and cities. Behind each map layer is a table containing an extensive database of information about each feature in that layer. By carefully analyzing these data, you can identify patterns in the data that are difficult to discover through visual examination alone.

Planning to learn

Each unit of the series leads you through a well-tested learning process that builds upon your existing knowledge. Each unit begins with a warm-up exercise designed to stimulate your thinking about the major concepts presented in the unit and the key questions that motivate and guide scientific research. It will help you frame your own questions about the topic—questions that you may be able to answer for yourself as you learn more in later investigations.

In the first investigation, you will explore maps and data looking for patterns. As you examine these patterns, you should ask yourself questions such as:

- Where do they occur? (or not occur?)
- Why does this pattern occur here and not elsewhere?
- What might cause this pattern?
- What else is spatially associated with this phenomenon?
- Do these things usually occur together in the same places?
- How has this pattern changed spatially through time?

A brief reading provides key background information about scientific principles and concepts, and should help you begin to answer the questions raised earlier.

Finally, in one or more additional investigations, you apply your new knowledge to solve a particular problem. This helps you measure your understanding of the material and apply the concepts you have learned to a new location or situation.

GIS made easier

The purpose of these investigations is not simply to learn how to use GIS, but to use one as a tool to explore and learn about natural processes and features and how they relate to humans and human activities. For this reason, all of the data have been assembled into ready-to-use projects, and complex operations have been eliminated or simplified. Although it is helpful for you to have basic computer skills, you do not need experience with ArcGIS software to complete the investigations. The ArcMap user interface has been modified to simplify complex and repetitive processes. Directions for each task are provided in the text, so you will learn to use the tool as you explore with it. The investigations barely scratch the surface of the data that have been provided, and we encourage you to explore the data on your own.

Instructional Design: The 5-E Learning Cycle

This series was designed using the 5-E learning cycle model, which promotes inquiry and exploration as a process for learning science. The Learning Cycle, originally credited to Karplus and Thier (*The Science Teacher*, 1967) and later modified by Roger Bybee for the Biological Sciences Curriculum Study (BSCS) project, proposes that learning something new or understanding something familiar in greater depth involves making sense of both prior experience and firsthand knowledge gained from new explorations. The 5-E model divides learning experiences into five stages: Engage, Explore, Explain, Elaborate, and Evaluate. Each stage builds on the previous stages as you construct new understanding and develop new skills.

The 5-E Learning Cycle

Engage (Warm-Up)

This stage is designed to help you understand the learning task and make connections to past and present learning experiences. It should stimulate your interest and prompt you to ask your own questions about the topic.

Explore (Investigation)

In this stage, you investigate key concepts by exploring scientific, geographic, and economic data sets. You begin identifying patterns in the data and connecting them to Earth processes. This further stimulates curiosity and new questions develop. You may diverge from the written investigation to explore your own questions, continually building on your knowledge base. Through this process of questioning and exploration, you begin to formulate your understanding of basic concepts.

Explain (Reading)

This stage introduces you more formally to the lesson's scientific and geographic concepts. You should gain a better understanding of major concepts, acquire important terminology, and verify answers to questions or problems posed earlier. In addition, more abstract concepts not easily explored in earlier activities are introduced and explained.

Elaborate (Investigation)

In the Elaborate stage, you will expand on what you have learned and apply your newfound knowledge to different situations. You will test ideas more thoroughly and explore deeper relationships.

Evaluate (Wrap-up)

At the end of each unit, you will use your understanding of key concepts to propose explanations and solutions to local or regional problems.

Learning science with GIS

Geographic Information Systems (GIS) provide an ideal vehicle for learning topics in Earth and environmental sciences and helping you develop scientific problem-solving skills. Formerly limited to professionals with access to high-end computer workstations, today GIS is accessible to many, and is being used by students from elementary through graduate school. GIS has a number of advantages over traditional materials when used as an instructional tool. These include:

- **Data visualization**—GIS-based investigations allow you to identify and characterize relationships by manipulating multiple visual representations of data, including dynamic and customizable maps, tables, charts, and animations.
- **Data analysis**—Analytical tools enable you to quantify relationships within and among spatial data sets using database functions, statistical analyses, and spatial overlay operations.
- **Multimedia integration**—Other forms of digital information, including animations, video, audio, and digital stills, can be woven into GIS activities, greatly enriching and extending your learning experience.
- **Technology literacy and transferable skills**—The use of GIS promotes technology literacy and provides you with skills transferable to your own research, other course work, and the workplace.

GIS-based instructional materials have the potential to enhance your learning by reinforcing concepts through discovery and by improving problem solving, visualization, and computational skills.

Tear out this sheet for easy reference to common ArcMap operations.

GIS Investigations for the Earth Sciences

Quick Reference Sheet

The ArcMap User Interface

Title Bar
Shows the name of the current map file.

SAGUARO Tools
Tools provided by the SAGUARO Project for these materials. Note: This toolbar may appear in a different location.

Menu Bar
Provides menus for performing various operations.

Toolbars
Tools for manipulating and analyzing map data.

Table of Contents
A list of data frames and map layers, and controls for changing the map view.

Status Bar
Displays tool descriptions, measurement results, and other information about the current operation.

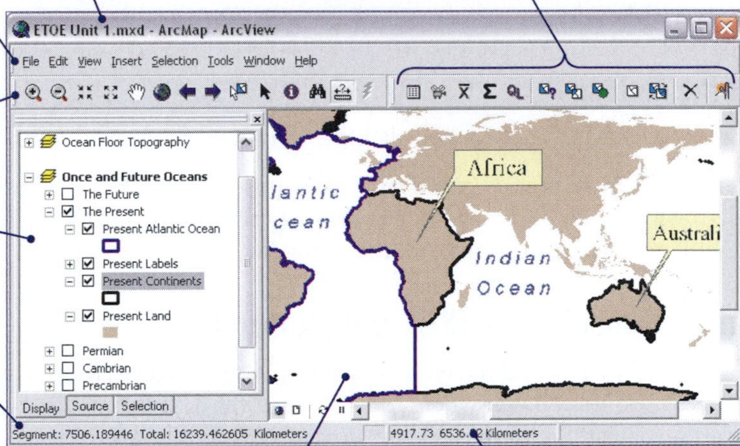

Data Frame
The currently-selected map.

Coordinates
Displays the coordinates, in the currently-selected units, of the cursor position.

Basic display operations

Activating a data frame

To activate a data frame, right-click its name and choose Activate from the pop-up menu. The title of the activated data frame is highlighted bold.

Expanding and collapsing data frames

To expand a data frame and show its layers, click the expand box ⊞. To collapse a data frame and hide its layers, click the collapse box ⊟.

Selecting layers or layer groups

To select a layer or layer group, click the layer or layer group name. Selected layers or layer groups are highlighted. To select multiple layers or layer groups, hold down the control key while clicking additional names.

Expanding and collapsing layers or layer groups

To expand a layer or layer group, click the expand box ⊞. To collapse a layer or layer group, click the collapse box ⊟.

Turning layers or layer groups on and off

To turn a layer or layer group on, check ☑ the box in front of its name. To turn a layer or layer group off, uncheck the box ☐ in front of its name. If a layer is turned on but is not visible, it may be hidden behind another layer. Try turning off the layers *above* that layer in the Table of Contents.

Table of Contents

Data Frames
Layer Groups
Layers

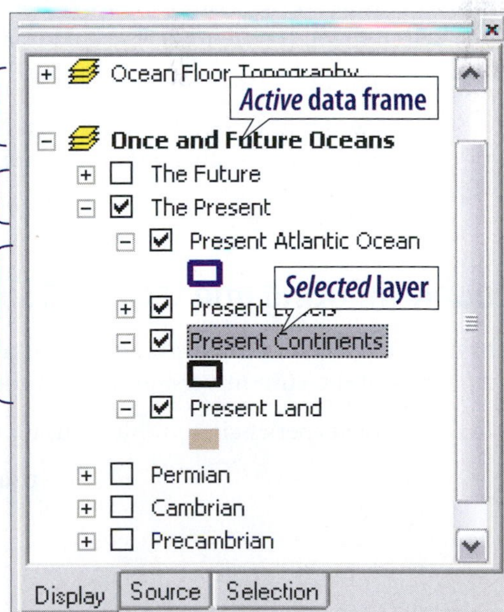

Opening ArcMap document (.mxd) files

At the beginning of each investigation, you will launch the ArcMap application and use it to open an ArcMap document file.

The instruction will look like this:

Launch ArcMap, and locate and open the **etde_unit_1.mxd** file.

etde_unit_1

ArcMap document files end with an .mxd file extension. Depending on how your computer has been set up, file extensions may not be visible, so the above file may appear as simply **etde_unit_1**. (Note: We will also refer to ArcMap document files simply as *map files*.)

Launching ArcMap

To launch the ArcMap application, click the Windows Start menu and choose **All Programs** > **ArcGIS** > **ArcMap**.

Shortcut to ArcMap.exe

Note: If there is an ArcMap shortcut icon on the desktop, Start menu, or taskbar, you can also use it to launch ArcMap. It may or may not show the .exe file extension. Your instructor or lab administrator may provide you with alternate directions for launching ArcMap.

Locating and opening the EXXX_Unit_X.mxd file

Choose **File** > **Open…** and navigate to where the specified ArcMap document file is located. Your instructor or lab administrator should tell you where the investigation data files are located. They will be in folders named with the module abbreviation and unit number, such as ETDE_Unit_1. Remember, you may not see the .mxd file extension.

Important note: Do not call or e-mail technical support if you cannot find an ArcMap document file. Your instructor or lab administrator can tell you where to find it!

Zooming

ArcMap has tools for zooming—enlarging and reducing areas of the map—that work like the tools you have used in other applications.

- To zoom in on an area, click and drag diagonally with the Zoom In tool to outline the area on the map. When you release the button, the area you selected will rescale to fill the data frame window.
- To zoom out, click anywhere on the map with the Zoom Out tool.
- If you zoom in or out so far that you do not know where you are, undo previous zooms by clicking the Previous Extent button.
- To view the entire data frame, click the Full Extent button.

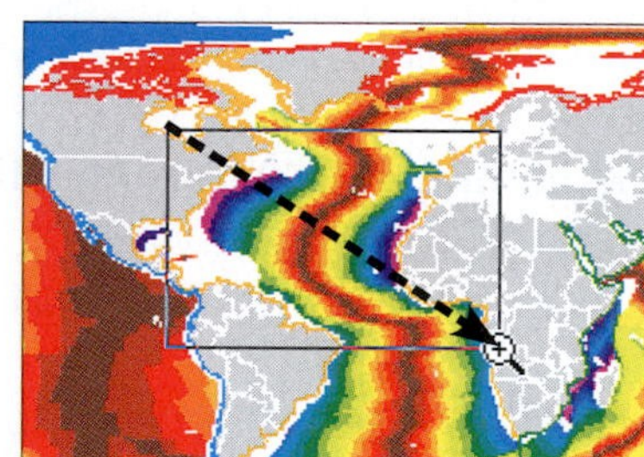

Troubleshooting and technical support

Your instructor or lab administrator may be able to help with common problems. If necessary, they can provide you with a copy of the current version of the troubleshooting guide for this series, **Troubleshooting.pdf**.

If you continue experiencing problems using these materials, contact the Thomson Learning technical support team at:

tl.support@thomson.com *or* **1-800-423-0563**

Unit 1
Searching for Evidence

In this unit, you will

- ***Investigate patterns in seafloor topography, earthquakes, and volcanoes.***
- ***Examine plate motion data.***
- ***Use these patterns and data to identify and classify plate boundaries.***

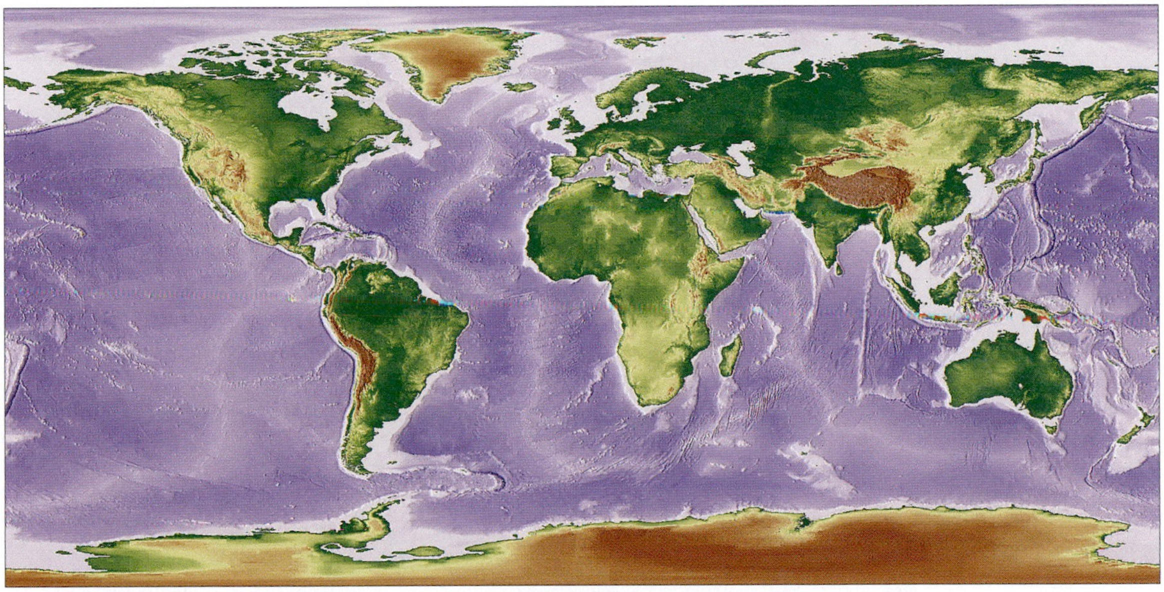

This ArcMap data frame showing global topography provides important evidence for the processes shaping Earth's crust.

Warm-up 1.1

A changing Earth

As a class, or in small groups as assigned by your instructor, discuss these questions.

1. The map on the first page of this unit shows that Earth does not have a smooth surface. What do you think causes variations in Earth's surface?

2. Is Earth changing? List any observable evidence you can think of that shows that the surface of our planet is or is not changing. The evidence can include both physical features and processes.

3. Try to think of changes to Earth's surface that might occur over

   a. short periods (100 years or less) —

   b. long periods (thousands to millions of years) —

In later activities, you will explore data to look for patterns in topography (the shape of the land), volcanoes, earthquakes, and the ages of seafloor rocks. These will provide important clues about how Earth's surface formed.

Investigation 1.2

Investigating Earth's clues

In this activity, you will explore patterns of features and events on Earth's surface. You will look for clues to help you understand how these patterns are interrelated and what they tell about the history of our planet. Your main tool in these investigations is a computer application called ArcMap GIS, a Geographic Information System.

- Launch ArcMap, and locate and open the **etde_unit_1.mxd** file.

Refer to the tear-out Quick Reference Sheet located in the introduction to this module for GIS definitions and instructions on how to perform tasks.

- In the Table of Contents, right-click the **Clues** data frame and choose Activate.
- Expand the **Clues** data frame.

Topography tales

Earth's story is told through its surface features, or *topography*. If Earth were made of only one kind of material, there would be no dry land. Earth's surface would be smooth and covered by a single global ocean. Fortunately, our planet is made of many types of rock, each with a different density. As gravity pulls the denser rock toward Earth's center, the less dense rock is pushed toward the surface. The result is a broken, bumpy surface with magnificent mountains of low density rocks and deep ocean basins of higher density rocks. You will begin this investigation by looking at Earth's topography in detail.

Inside Earth

Earth is composed of distinct layers with the densest material in the core and the least dense material in the thin crust.

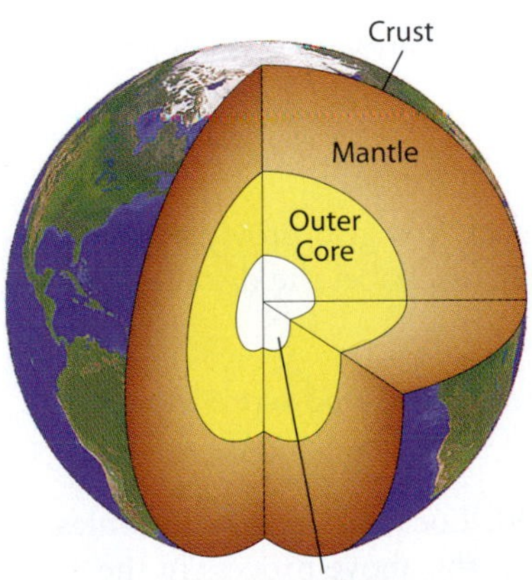

The **Clues** data frame shows a map of Earth's topography. The different colors represent different elevations.

1. In the elevation scale below the map on the screen (or to its right), what color is each of the following?

   a. The lowest topography (elevation) —

   b. Sea level —

   c. The highest topography —

The highest highs

The obvious place to look for evidence of change is where the topography is most extreme — its highest and lowest places. You will start your search by looking for Earth's highest places.

- Turn on the **Topographic Profiles** layer.
- Select the **Topographic Profiles** layer.

Still can't picture it?

To better understand what topographic profiles represent, click the Media Viewer button and choose **Topographic Profiles** from the media list.

Profile location map

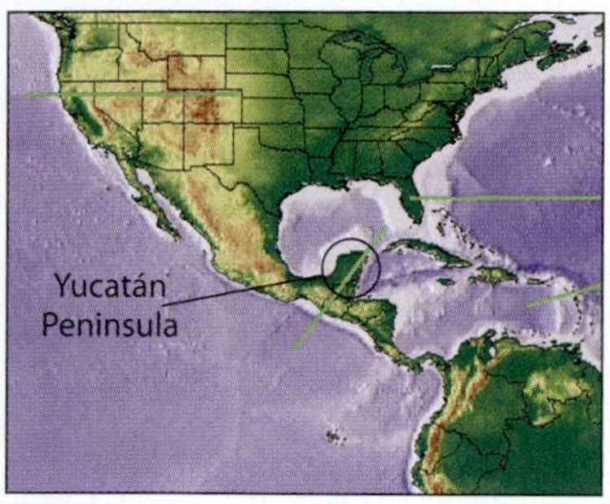

The lines in this layer mark the locations of topographic profiles, or cross-sectional views of the land surface. Imagine slicing through the Earth, pulling the halves apart, and viewing one half from the side. The result is a topographic profile. For example, the topographic profile below shows a slice through Mexico's Yucatán Peninsula. To make the land features easier to see, topographic profiles are usually "stretched" vertically—a process called vertical exaggeration.

Yucatán Peninsula

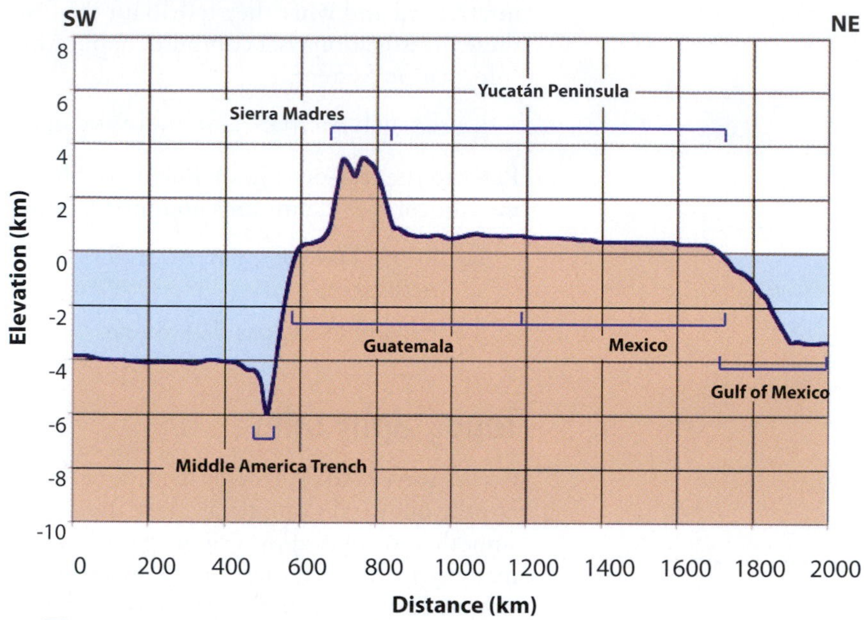

Examine the **Clues** data frame on your screen. Using the map elevation scale at the bottom or to the right of the map, find a place where a profile line crosses an area 3 kilometers or higher in elevation.

Next, you will sketch the topographic profile for this location and use it to estimate the highest elevation along the profile.

Using the Hyperlink tool, click the profile line you chose.

2. Sketch and label the topographic profile in the first grid provided on the following page. Include all geographic labels, and label the distance (in kilometers) on the horizontal (X) axis.

3. Estimate the highest elevation along the profile line, and write it in the space provided above the grid.

Close the profile plot window when finished. Locate three more profiles that cross areas higher than 3 km, and repeat the above process in the other three grids provided.

Location ______________________________

Maximum elevation = __________ *km*

Location ______________________________

Maximum elevation = __________ *km*

Location ______________________________

Maximum elevation = __________ *km*

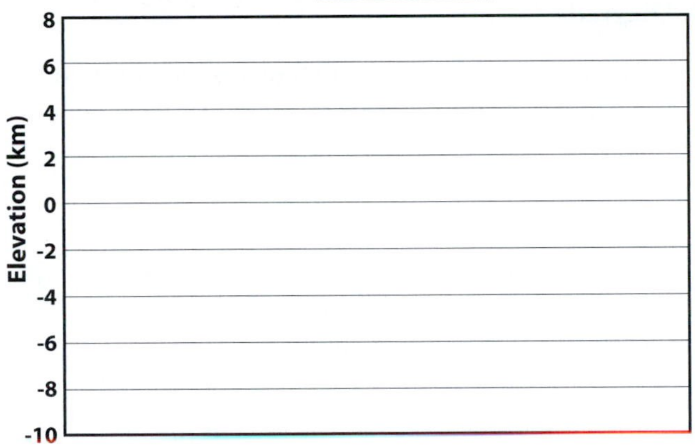

Location ______________________________

Maximum elevation = __________ *km*

Block diagrams

Think of block diagrams as pieces cut from the middle of a large layer cake and viewed at an angle. Notice that the upper edge of each face of the block is a topographic profile. In addition to surface features, block diagrams usually show layers of rock and structures below the surface.

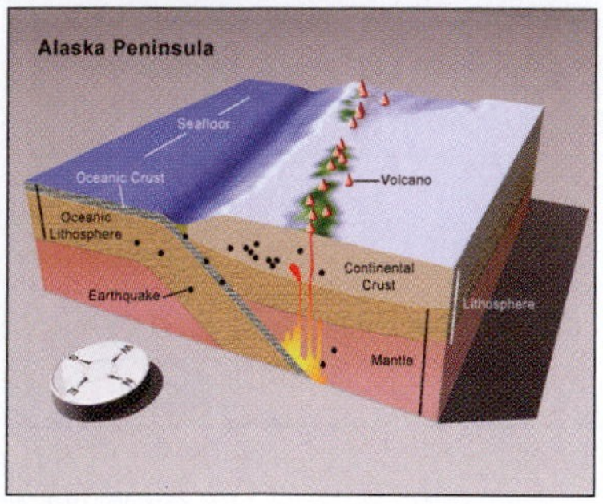

Not-so-green Greenland & not-so-icy Iceland

Place names can be deceiving. Although there's some green land along the coasts of Greenland, 85 percent of the land is covered by a continental ice sheet up to 3 km thick. This represents 10% of the world's total freshwater reserves.

In contrast, only 11 percent of neighboring Iceland is covered with glaciers and ice fields, and these are much thinner than Greenland's ice sheet.

Like icing on a cake

Greenland and Antarctica are two of Earth's highest regions. Many of their elevations, however, are not rock but another material. (Remember, the brown color in the shaded relief image represents high elevation, not the surface material.) To compare Greenland and Antarctica with other high regions and find out what this "other material" is, you will examine block diagrams of the areas.

- Turn off the **Topographic Profiles** layer.
- Turn on the **Block Diagrams** layer.
- Select the **Block Diagrams** layer.
- Using the Hyperlink tool, click the profile lines crossing Greenland, Antarctica, and any other high regions you identified above, to display a block diagram of each location. Close each block diagram window when you are finished viewing it.

4. In the block diagrams, what is the top layer of surface material that adds as much as 1500 meters (5000 feet) to the elevation of Greenland and Antarctica?

- Turn off the **Block Diagrams** layer.

For this reason, it's probably best to ignore these two high places when thinking about the connection between topography and Earth's changing surface. Still, even these places fit the pattern of less dense material forming higher topography.

The ocean's unexpected highs

Before maps of seafloor topography became available, most scientists believed the seafloor was relatively flat. Personal experience with lakes and rivers often suggests that the deepest part of the ocean should be near the middle.

5. Use the elevation scale to examine the ocean basins. Are they smooth bottomed? Does their depth increase continuously toward the middle? If not, describe the features you do find there. (Are these features like plains, like valleys, or like mountains?)

- Click the QuickLoad button.
- Select **Spatial Bookmarks**, choose **Atlantic Ocean Middle** from the list, and click **OK**.

The seafloor in the middle of the Atlantic Ocean should now be centered on the map.

- Turn on the **Topographic Profiles** layer.
- Select the **Topographic Profiles** layer.
- Using the Hyperlink tool, click on the profile line that crosses that region.

6. Sketch the topography along the profile line between Florida and Africa on the graph below. Label the distance on the horizontal (X) axis.

Click the Full Extent button to view the entire map again.

Features like the one running down the middle of the Atlantic Ocean are called *spreading ridges*. They look like long, jagged scars and in a sense they are. To better understand the topography of the Mid-Atlantic Ridge, look at the three topographic profiles that cross it.

Using the Hyperlink tool, click on each of the profile lines that cross the ridge and examine its profile. These are located at about 23° S, 28° N, and 44° N latitude. (You may turn on the **Latitude / Longitude** layer and use the Identify tool to click on the latitude and longitude lines as needed to determine their locations. Be sure to select **Latitude / Longitude** from the list of layers in Identify Results window.) Close each profile window when you are finished viewing it.

7. Where else on the map do you see spreading ridges besides the Atlantic Ocean? (Turn on the **Ocean Labels** layer if you need help identifying the oceans.)

Spreading ridge

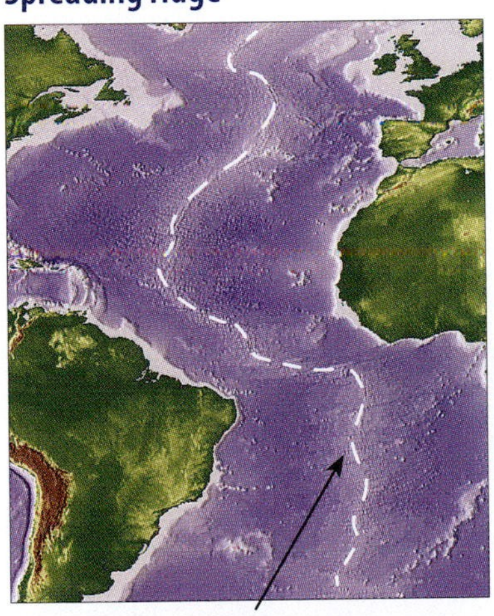

Shallow linear ridge in Mid-Atlantic Ocean

The lowest lows

Earth's lowest spots aren't in the middle of the oceans, so they must be elsewhere in the ocean basins.

Using the Hyperlink tool, click on the profile line that crosses the west coast of South America to see the topography of this area.

8. Sketch the topographic profile on the graph on the following page. Label the deepest point along this profile with its estimated depth. Then label the distance on the horizontal (X) axis.

These deep linear features, the lowest points on Earth, are called *trenches.*

9. About how far from the edge of the continent is the deepest part of the trench? (Estimate the distance from the topographic profile.)

Close the topographic profile window.

Use the elevation scale and the Hyperlink tool to examine other profiles, and identify two other locations where you find trenches similar to the one near the west coast of South America.

10. From the profiles, give the name and estimated depth of each trench along its profile.

 a. Name— Maximum depth—

 b. Name— Maximum depth—

The deepest points on Earth

Each profile drawn in the **Topographic Profiles** layer does not necessarily cross the deepest part of the trench along its length. The maximum depths (km) recorded for each trench are listed below.

Trench Name	Maximum Depth (km)
Aleutian	8.1
Atacama	8.1
Java	7.7
Kuril	10.5
Mariana	11.0
Middle America	6.6
Puerto Rico	8.6
South Sandwich	8.3

The lowest point on Earth is called Challenger Deep, part of the Mariana Trench just beyond the eastern boundary of the Philippine Sea. It reaches a depth of over 11 km below sea level. The highest point on Earth is Mt. Everest in Nepal, nearly 9 km above sea level. That means the range in Earth's topography is over 20 km.

Do you get the picture?

Now it's time to see how well you can interpret Earth's topography. A profile line is shown on the map below.

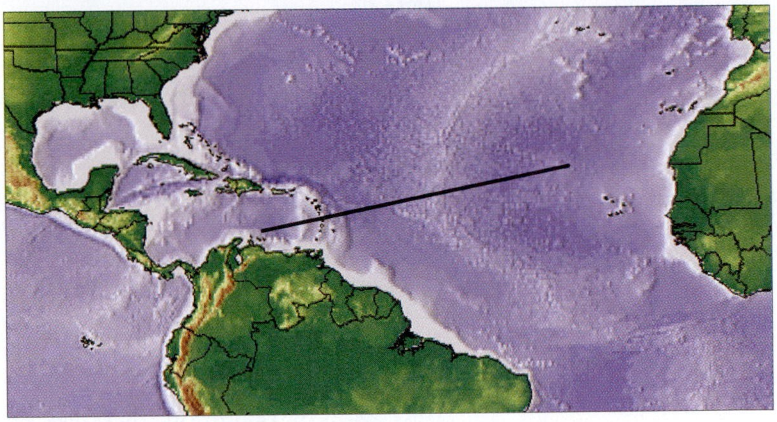

11. Without looking at the profiles on your map, sketch what you think the profile along the line above would look like. Label any ridges and trenches on your profile. Estimate the distance (in kilometers) along the horizontal (X) axis.

Volcanic clues

The discovery of spreading ridges made scientists wonder what caused these long features to form underwater in the middle of the ocean. You will investigate these next.

- Turn off all layers except **States**, **Countries**, and **Topography**.
- Turn on the **Ridge Volcanics** layer.

Ridge volcanics

Since discovering spreading ridges in the early 1960s, scientists have studied them extensively using sonar and underwater research vessels. They have learned that spreading ridges form where rising magma — molten rock from Earth's mantle — breaks the ocean floor along long cracks, or fissures. When a fissure opens, magma squeezes up through it and solidifies, adding to the rock on both sides of the ridge and filling in the crack. Generally, the magma does not create individual volcanoes or lava flows at the ridges, and the relatively gentle underwater eruptions go unnoticed. The **Ridge Volcanics** layer shows the locations of the fissures, which run down the center of the ridges.

- To see an animation of a ridge volcanic eruption, click the Media Viewer button and choose **Ridge Flyby movie**. Close the Movie window when you have finished viewing the animation.

To better understand ridge volcanics, you will look at a block diagram of the Mid-Atlantic Ridge. Note the symmetrical appearance of the ridge.

- Turn on the **Block Diagrams** layer.
- Select the **Block Diagrams** layer.
- Using the Hyperlink tool, click on the profile line that crosses the Mid-Atlantic Ridge to view a block diagram of the ocean floor.

A spreading ridge is not built up layer by layer from volcanic eruptions. Rather, the ridge exists because the warmer rock near the center of the ridge is less dense than the cooler rock farther from the ridge. Buoyancy forces this warmer rock to rise higher, forming the ridge. As fresh, hot magma is injected into the fissures, the older rock moves away from the ridge and the heat source. As it moves away, it cools, becomes denser, and sinks to form the deep-ocean floor.

- Close the Block Diagram window.
- Turn off the **Block Diagrams** layer.

Ridge volcanics aren't limited to large ocean basins.

12. Find two places where ridge volcanics lie in narrow seas between continental land masses. Circle their locations on the map on the next page. (Hint: Use the Zoom In tool to take a closer look.)

Features of spreading ridges

Recent research has provided a picture of the unique environments of spreading ridges. Among the unusual features of these ridges are hydrothermal vents of superheated water called "black smokers." These undersea hot springs circulate hot, mineral-rich water and dissolved volcanic gases that support a variety of life.

NOAA/PMEL

A black smoker vent on the southern Juan de Fuca Ridge off the Oregon coast. To see a movie of this black smoker, click the Media Viewer button and choose **Black Smoker** from the media list.

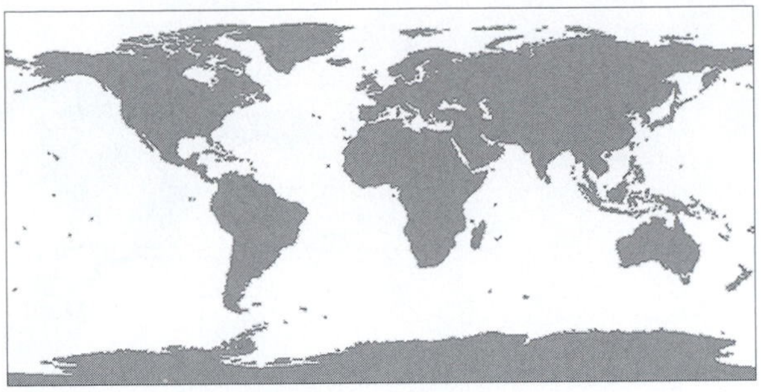

Trenches and volcanic arcs

Turn on the **Trenches** layer.

The **Trenches** layer shows the locations of the deep-ocean trenches you learned about earlier in this activity. These trenches are the deepest parts of the ocean basins.

13. Are trenches found closer to land or closer to the mid-ocean?

Turn on the **Volcanoes** layer.

In the **Volcanoes** layer, each brown triangle represents an individual volcanic feature that has been active within the past 10,000 years.

14. Ridge volcanics typically occur in the ocean. Where are the majority of individual volcanoes found?

Turn off the **Ridge Volcanics** layer.

There are two fairly distinct populations of volcanoes around the world — those in linear arrays or clusters and those that are isolated from one another.

15. Are the clusters of volcanoes typically located near trenches, near the edges of the continents, or both? (Hint: Use the Zoom In tool to take a closer look.)

Look closely at your map to find several of these clusters of volcanoes.

16. On the map on the following page, circle the locations of four volcano clusters.

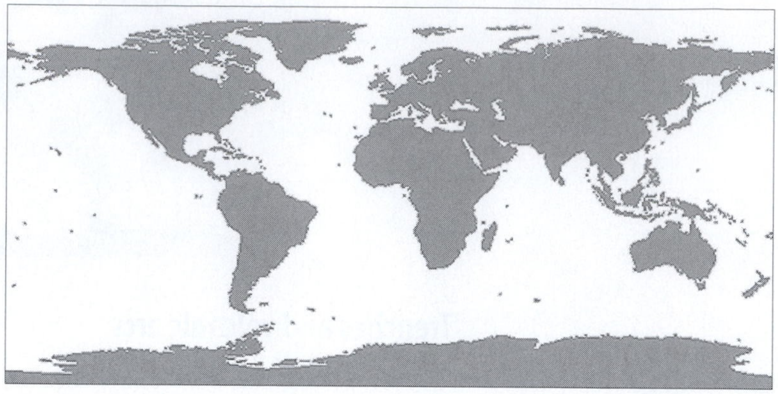

Many of the isolated volcanoes are created by *hot spots*. These are hot plumes of magma that rise toward the surface from deep within the mantle. Later you will learn about the processes that create these two types of volcanoes and about differences in their characteristics. For now, it is important to know the two types exist.

Volcanoes and your community (optional)

Up to now, you have been looking at the "big picture" of volcanoes. Now you will learn about the volcanic feature nearest to your community.

- Click the Full Extent button to view the whole map.
- Use the Zoom In tool to zoom in on the area in which you live.
- Select the **Volcanoes** layer.
- Using the Measure tool, click on your community and drag the cursor to the volcanic feature nearest to your community. When you have found the nearest feature, double-click on it. Read the distance in the status bar at the bottom left corner of the screen.

17. How far, in kilometers, is your community from the nearest volcanic feature?

- Click the Identify tool.
- In the Identify Results window, select the **Volcanoes** layer from the list of layers.
- Using the Identify tool, click on the volcanic feature you identified as nearest to your community to learn more about it.

18. Write the name (if any), location, date of last eruption, and any other information you learn about the volcanic feature nearest to your community.

- Close the Identify Results window.
- Click the Full Extent button to view the whole map.

"I know there's a volcano near my community, but it's not on the map!"

Remember, this map shows only the volcanoes that are known to have been active in the past 10,000 years.

"Nothing happens when I click with the Identify tool!"

If the Identify Results window does not open or displays incorrect information when you click on the volcano symbol, check these things:

- Is the Identify tool selected?
- Is the **Volcanoes** layer selected?
- Did you select **Volcanoes** from the list of layers in the Identify Results window?
- Are you clicking directly on the volcano symbol?

Seismic—refers to a vibration in the Earth caused by an earthquake or some other event, such as an explosion or meteorite impact.

Seismic clues

Previously, you looked for patterns in topography and the locations of volcanic features to understand Earth's changing surface. Now you will look for patterns of earthquakes around the world and explore their relationship to topography and volcanoes.

- Turn off all layers except **States**, **Countries**, and **Topography**.
- Turn on the **Latitude / Longitude** layer.

What is an earthquake?

Magnitude—a measure of the amount of energy released during an earthquake. The scale starts at 1, with each increasing level in magnitude being ten times greater than the previous level. For example, an earthquake of magnitude 5 is ten times more powerful than an earthquake of magnitude 4.

An earthquake is a vibration of Earth caused by a sudden release of energy, usually as an abrupt breaking of rock that results from the slow accumulation of stress in the rocks. However, only brittle, cold rocks can be broken in earthquakes. If rock is heated, it can stretch and deform slowly over time without breaking. The size or *magnitude* of an earthquake is determined in part by how much rock is broken in an event.

- Turn on the **Earthquake Depth** layer.
- Select the **Earthquake Depth** layer.

Earthquake depth

The data in this layer are displayed according to depth. The same data are displayed according to earthquake magnitude in the **Earthquake Magnitude** layer, which you will explore later.

This layer shows the locations of nearly 12,000 earthquakes recorded around the world since 1973. The data include all earthquakes of magnitude 5 and greater from 1973–2004, plus a few smaller events from regions that do not have large earthquakes.

Viewing the map as a globe

There are disadvantages to viewing Earth as a flat map. It's not obvious that a feature that disappears off the left edge of the map continues on the right edge, or that features at the top or bottom actually meet at the poles. The real Earth is nearly a sphere, and the earthquake patterns you are looking for may be easier to see if you view Earth as a globe.

- Launch the Google Earth application.
- Double-click the **Earthquake Depth.kmz** file located in the etde_unit_1 folder. This file should open within the Google Earth application.

Now you are viewing the earthquakes on a globe. Use the **Earthquake Depth** legend in ArcMap to examine the earthquakes in Google Earth.

- Left-click and drag the cursor on the globe to turn it in any direction.

19. Describe any patterns you see in the distribution of earthquakes over Earth's surface. (For example, do they form lines, arcs, circles, or clusters? Are the patterns connected or disconnected?)

- Quit Google Earth, do not save changes, and return to ArcMap when you are finished.

Earthquake depth patterns

Next, you will look for patterns in the depth of earthquakes around the world.

Examine the legend of the **Earthquake Depth** layer in the Table of Contents.

20. What color are the *shallowest* earthquakes? What color are the *deepest* earthquakes?

    a. Shallowest —

    b. Deepest —

Turn on the **Trenches** and **Ridge Volcanics** layers.

Use the Zoom In tool to zoom in on South America, and look at the ridge and trench that are located near the west coast of that continent.

Look closely at the ridges and trenches. The earthquake depth patterns at these two types of features are different.

21. How deep are the earthquakes near each type of feature? (Give a range, from the shallowest to the deepest.)

    a. Ridges —

    b. Trenches —

Next, you will look at differences in earthquake depth around the world.

Click the Full Extent button to view the whole map.

Use the Zoom In tool and the Pan tool to look closely for regions with deep earthquakes.

22. Are there any earthquakes deeper than 150 km near spreading ridges? If not, why do you think this is?

Where am I?

If you need help with the names of countries for question 23, turn on and select the **Countries** layer and use the Identify tool to click on a country to read its name. (Be sure to select **Countries** from the list of layers in the Identify Results window.)

23. Identify three areas with many earthquakes deeper than 150 km. (For example, "The west coast of Central America near Mexico, El Salvador, and Nicaragua.")

- Use the Zoom In tool to closely examine the pattern of the earthquakes in the areas you identified in the previous question.

24. In general, how do earthquake depths change as they occur farther away from the trench?

- Turn on the **Volcanoes** layer.
- Click the Full Extent button to view the whole map.
- Turn on the **Block Diagrams** layer.
- Select the **Block Diagrams** layer.
- Using the Hyperlink tool, click on each of the profile lines that crosses a trench to view a block diagram at that location.

25. What do you think is happening at the trenches to cause the deep earthquakes there? Why don't many deep earthquakes occur elsewhere?

- Turn off the **Earthquake Depth**, **Latitude / Longitude**, **Block Diagrams**, and **Volcanoes** layers.

Earthquake magnitude patterns

Next, you will look for patterns in earthquake magnitudes around the world.

- Turn on the **Earthquake Magnitude** layer.
- Select the **Earthquake Magnitude** layer.
- Examine the **Earthquake Magnitude** legend in the Table of Contents.
- Use the Zoom In tool and the Pan tool to look more closely for regions with large earthquakes.

26. Where do you see more strong earthquakes (magnitude 7 or greater) —near the spreading ridges or near the deep trenches?

- Click the Full Extent button to view the entire map.

Comparing depth and magnitude

To see if there is a relationship between the magnitude and depth of earthquakes, you will examine only the strongest and deepest earthquakes and look for a connection between them.

- Turn off the **Earthquake Magnitude** layer.
- Turn on the **Large Earthquakes (mag 7+)** layer.

You will now see only those earthquakes that are magnitude 7 or greater.

Turn on the **Deep Earthquakes (> 150 km)** layer.

The map should now show earthquakes that occurred at depths of 150 km or deeper, in addition to the strong earthquakes.

27. Compare the locations of strong earthquakes (magnitude 7 or greater) to the locations of deep earthquakes (depths 150 km or greater). Do they occur together, or are they found in different regions? Explain.

28. Spreading ridges were formed recently from hot magma. How might this affect the magnitude and depth of earthquakes near the ridges?

Quit ArcMap and do not save changes.

Reading 1.3

Earth's inner structure

In addition to differences in composition, Earth's inner structure is defined by differences in the state of the material. The interior of Earth is still hot enough for the outer core to be molten and for the mantle to flow like putty.

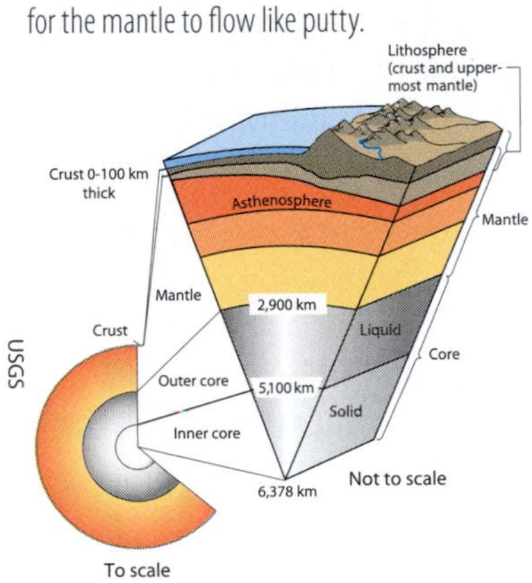

Discovering plate tectonics

The theory of **plate tectonics** states that Earth's outermost layer is fragmented into a dozen or more large and small plates that are moving relative to one another. The **lithosphere**, comprised of plates, is a rigid layer made up of the crust and uppermost rigid part of the mantle. The plates ride on top of the hotter, more mobile material in the deeper mantle.

Heat and gravity are the two main driving forces of plate tectonics. Earth's core reaches temperatures over 6000 °C. About half of this heat comes from the decay of naturally occurring radioactive minerals. The other half is left over from the heat generated when Earth formed. Heat energy travels slowly from the interior to the surface by the processes of convection and conduction.

In conduction, heat energy moves by collisions between atoms or molecules. Convection occurs as rock heats up, expands, and becomes less dense. Gravity forces this buoyant rock to rise toward the surface. There it cools, becomes denser, and eventually sinks deep into the mantle again. Convection cools Earth more efficiently than conduction. Still, it takes a long time to cool an object the size of Earth.

At mid-ocean ridges, hot rock rising from deep in the mantle forms new oceanic lithosphere. This lithosphere floats about 2 km above the average depth of the ocean basins. Gravitational forces push the plates down and away from this high ridge. At trenches, old, cold, and therefore dense lithosphere dives into the mantle due to gravitational pull. The resulting ridge-push and trench-pull forces created by heat and gravity drive plate tectonics.

Evidence of change

Earthquakes occur when forces, or **stresses,** build up in cool, brittle rock and are released suddenly as the rock breaks. These stresses are often found at the edges of plates, where the plates move together, pull apart, or grind past one another.

Earthquakes—reshaping the land

A major factor controlling where earthquakes occur and the amount of energy released is rock strength. Rock strength depends on the temperature and the rate at which the rock is deformed. Rocks near the surface are cool and brittle. When stressed beyond their mechanical strength, they break like a rigid pencil, sending out shock waves in all directions. Typically, rocks below 20–30 km depth are warmer and more ductile than the shallower rocks. When they are stressed, they deform rather than break.

Earthquakes often cause changes in topography. Over millions of years, they can produce great mountain ranges and large valleys.

Volcanoes—creating new crust

Volcanoes are another indicator of changes occurring in Earth's crust due to plate tectonics. Volcanoes develop where molten rock rises from deep inside Earth to the surface, then cools and forms new lithosphere.

Globe-circling "wound"

American oceanographer Bruce C. Heezen described the mid-ocean ridge system as "the wound that never heals." Hidden beneath the ocean surface, these ridges extend for thousands of kilometers forming an interconnected network that encircles Earth.

Most of the world's volcanic activity occurs along the spreading ridges found throughout the world's ocean basins. At spreading ridges, hot rock rises to form new oceanic lithosphere. As the magma rises beneath the ridge, the plates pull apart at the cracks, or **fissures**, running down the center of the ridge. Magma squeezes into the fissure and cools, solidifying onto the edges of the two plates. Gravity pulls the plates slowly down and away from the ridge's center like a conveyor belt.

Periods of volcanic activity along the ridges are separated by quiet intervals with no activity. On human time scales, the process appears irregular, but on geologic time scales of millions of years, it is constant. On average, most spreading ridges extend between zero and tens of centimeters per year.

Hot-spot volcanism

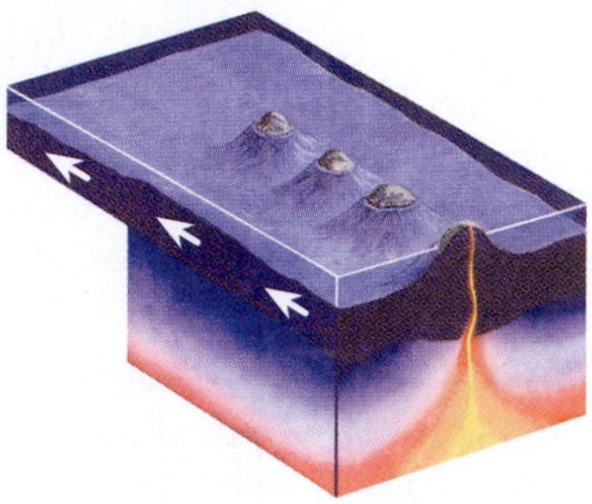

USGS

Hot-spot volcanoes

Another type of volcano forms above **hot spots**, or hot plumes of magma that rise from deep within Earth's mantle and melt through the crust. As plates move over the stationary hot spots, the magma plume continues to melt through the crust, creating chains of volcanoes. The ocean floor has many chains of hot-spot volcanoes, like those found in Hawaii and Iceland. Though well known, hot spots like these create the least common type of volcano.

Subduction zones—recycling crust

Gravity simply will not allow Earth to expand like a balloon. Thus, when new lithosphere is created at spreading ridges, it must either be compressed or recycled somewhere else. Which process occurs depends on the lithosphere's density. High-density oceanic lithosphere is recycled, whereas low-density continental lithosphere is compressed.

Subduction zone

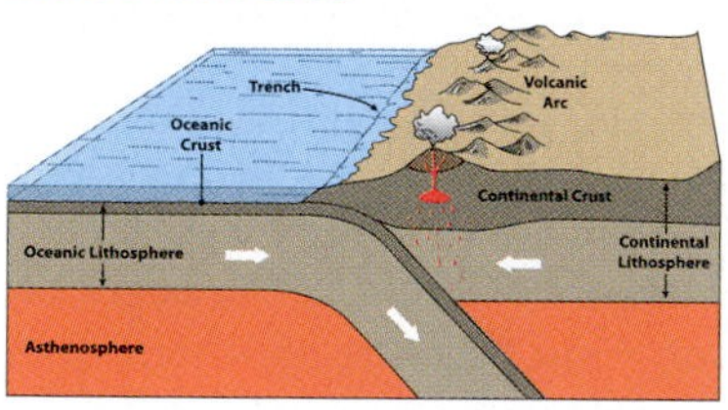

USGS

A subduction zone occurs at a convergent plate boundary, where a denser lithospheric plate plunges beneath a less dense lithospheric plate.

Oceanic lithosphere is recycled into the mantle in a process called **subduction**. At subduction zones, gravity pulls the denser oceanic lithosphere deep into the mantle, forming trenches in the ocean floor up to 11 km deep. Because of density differences, trenches often develop near boundaries between oceanic and continental lithosphere.

As oceanic lithosphere is subducted, it is warmed by the surrounding mantle rock. At a certain depth, water trapped in the minerals of the subducting plate is released due to pressure, causing the surrounding rock to melt. The molten rock rises, melts through the overriding plate, and creates a chain of volcanoes on the surface called a **volcanic arc**. The Cascade Mountains in the Pacific Northwest of the U.S. are an example of a volcanic arc.

Continental mountain ranges

When two continental plates collide, the low-density rocks are too buoyant to be subducted. Instead, they smash into one another, forming huge mountain ranges. For example, the Himalayas are being thrust upward by the collision between the Indian and the Asian plates. The Appalachian Mountains of the Eastern U.S. are an eroded remnant of a similar collision in the past. When first formed, they may have rivaled the Himalayas in size.

Classifying plate boundaries

Using earthquake, volcano, and topographic patterns around the world, scientists mapped the major plate boundaries. Once the boundaries were identified, they figured out how the plates interact and used this information to classify each boundary. There are three basic types of plate boundaries, identified according to the plate motions at the point of contact.

Divergent boundaries

At divergent boundaries (such as spreading ridges), plates are moving away from each other. The plates in these areas are relatively thin (less than about 30 km deep), weak, and warm; earthquakes here are generally shallow and small to moderate in size. As the plates move away from one another, the gap is filled with molten rock that forms a high ridge. The age of rock that was formed at divergent boundaries increases with distance from the boundary. Occasionally, divergent boundaries develop within continental plates, causing them to split apart. This is happening in East Africa today, and may eventually create a new ocean basin and spreading ridge there.

Convergent boundaries

At convergent boundaries, plates collide and deform. The most common type of convergent boundary is a **subduction zone**. A subduction zone is characterized by a strong, cold oceanic plate diving into the mantle and forming a deep trench along the entire boundary.

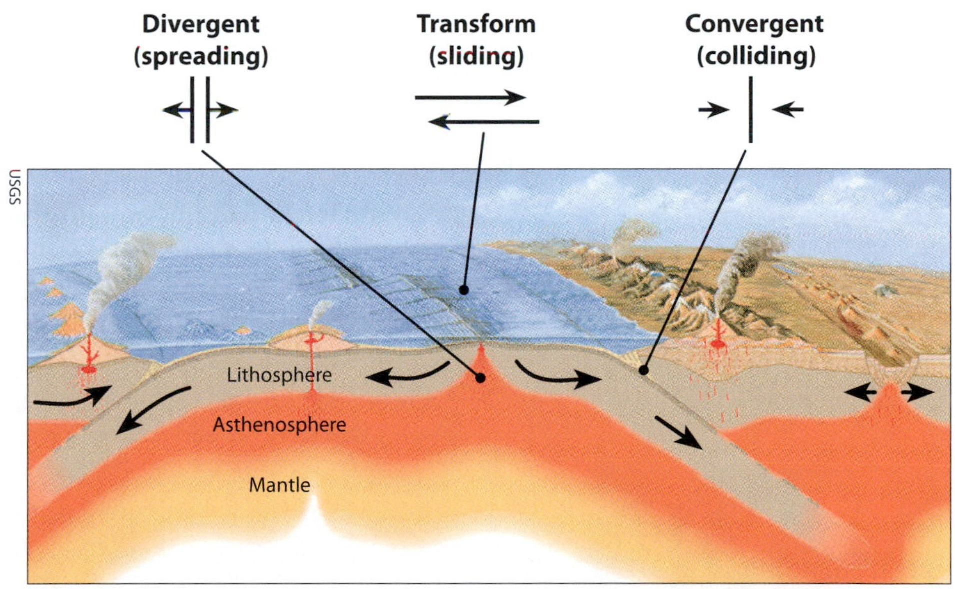

Types of plate collisions

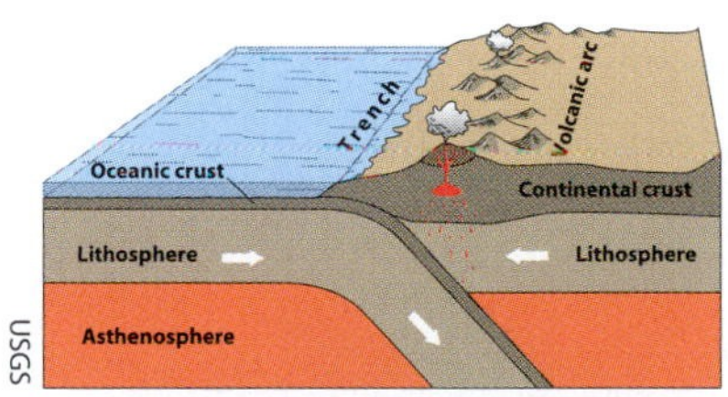

Oceanic–Continental collision

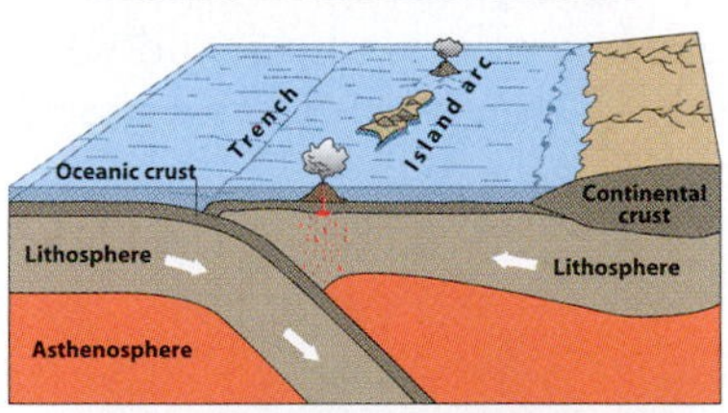

Oceanic–Oceanic collision

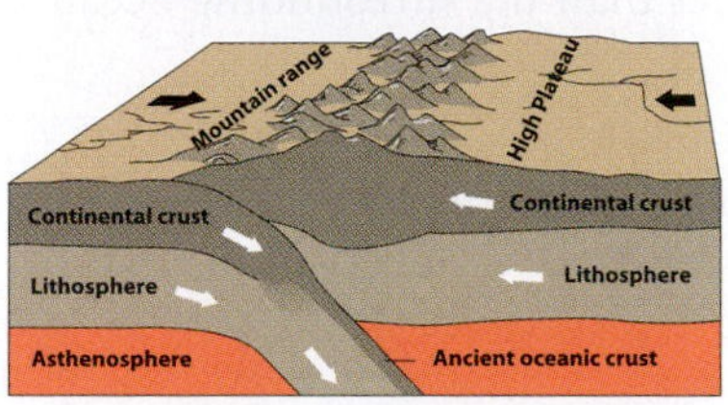

Continental–Continental collision

Continental plates are very buoyant and cannot be subducted. When continental and oceanic plates collide, the oceanic plate will always be subducted because it is more dense. Where two continents collide, the plates crumple and build large mountain ranges like the Himalayas. Oceanic plate gets older and becomes colder and more dense as it moves away from a spreading ridge. Thus, if two oceanic plates collide, the older, denser plate will subduct into the mantle.

Volcanoes are created along the plate boundary where the oceanic plate descends into the mantle, carrying with it minerals that contain water. As the descending plate reaches a depth near 150 km, these minerals release the water, which mixes with the mantle rock, allowing it to melt more easily. The lower-density molten rock melts its way upward from the mantle into the overriding plate, forming volcanoes where it reaches the surface. These volcanoes pose a hazard to humans because they are often very explosive and located near population centers. Mount St. Helens and Mount Fuji are examples of this type of volcano.

Earthquakes occur along the zone of contact between two converging plates and at great depths in the subducting material. Rock conducts heat very poorly, so the subducted material stays cold and rigid for hundreds of millions of years, and hundreds of kilometers deep. At subduction zones, earthquakes occur along the cold, brittle subducting plate. This pattern of shallow to deep earthquakes is an identifying characteristic of subduction zones. Most of the world's strongest earthquakes occur along subduction zones, where the contact area of rock against rock (or plate against plate) is both deep and wide.

How does water help rock melt?

Although it seems against common sense, mixing two substances often lowers the melting point of one or both substances. This is the case with rock, which melts at a lower temperature when water molecules are added to the mix.

Transform boundaries

Transform boundaries occur where the edges of two plates grind past each other without converging or diverging. The San Andreas Fault in California is a transform plate boundary. The linear features in the seafloor that are situated perpendicular to spreading ridges are also transform faults. At transform boundaries, the thickness of the lithosphere ranges from a few kilometers near ocean ridges to 100 – 150 km where they cut across continents.

Earthquakes occurring at transform boundaries in the seafloor can be very strong if the rupture occurs along a great length, but are usually moderate and of shallow depth (less than 30 km). The weaknesses created by the faulting may help provide routes for magma to reach the surface more easily, but volcanoes along these boundaries are not common.

Transform boundary

Robert E. Wallace, USGS

The San Andreas fault, a continental transform boundary, is visible in this aerial photo near San Luis Obispo, California.

Questions

1. What two forces drive the processes that change Earth's surface?

2. Knowing that the lithosphere "grows" at spreading ridges, why would we suspect that trenches and subduction zones exist, even if we can't see them?

3. Why is the center of a spreading ridge higher than the surrounding ocean basin?

Other evidence for plate tectonics

In Unit 2, you will investigate magnetic patterns in oceanic rocks that provide evidence of plate tectonics. Other evidence that you won't explore in this activity include paleoclimate and fossil evidence.

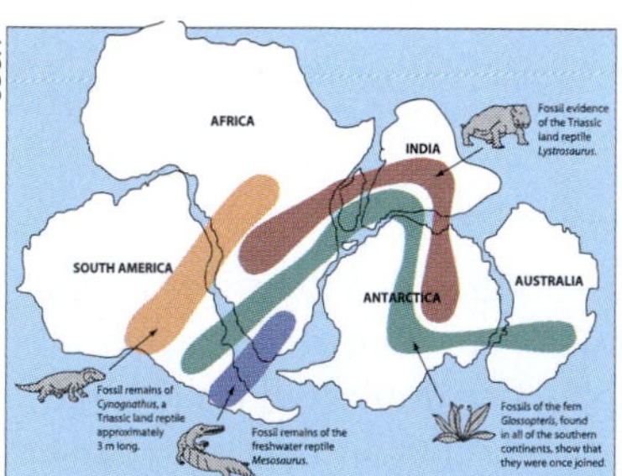

USGS

Fossil evidence shows that present-day continents of the Southern Hemisphere were once part of a larger supercontinent, now called Gondwanaland.

4. At what types of plate boundaries would you find a narrow band of shallow earthquakes?

5. At what types of boundaries would you find a wide band of earthquakes, extending to depths greater than 150 km?

6. Why are most subduction zones located along the edges of continents?

Investigation 1.4

Analyzing plate boundaries

You have looked at patterns in earthquakes, volcanoes, and global topography and learned how these are related to plate movements. Using all of this evidence, you will predict where you think the surface of Earth is deforming along the plate boundaries.

- Launch ArcMap, and locate and open the **etde_unit_1.mxd** file.

Refer to the tear-out Quick Reference Sheet located in the introduction to this module for GIS definitions and instructions on how to perform tasks.

- In the Table of Contents, right-click the **Clues** data frame and choose Activate.
- Expand the **Clues** data frame.

Locating plate boundaries

- Turn the **Trenches**, **Earthquake Depth**, **Volcanoes**, and **Ridge Volcanics** layers on or off to help you locate the plate boundaries.

1. Look at all the available evidence and use it to determine where you think Earth's plate boundaries are located. Draw the boundaries on the map on page 28. Use a solid line where you're sure of a boundary and a dashed line where you're not sure.
2. Add a legend to your map to show what your symbols mean.

Comparing to "accepted" plate boundaries

After you have drawn your map of plate boundaries, compare it to the currently accepted plate boundary map.

- Click the QuickLoad button.
- Select **Data Frames**, choose **Locating Boundaries** from the list, and click **OK**.

This data frame shows the locations of the accepted plate boundaries.

3. On your map, circle the areas where your boundaries are different from the accepted boundaries.
4. What do you think accounts for the differences between the plate boundaries you drew and the accepted plate boundaries?

There is disagreement within the scientific community on the location and existence of several plate boundaries. This is especially true in areas at the far northern and southern latitudes, where few instruments are available to monitor earthquakes and plate motion; and in the Indian Ocean and the Mediterranean, where the plates are moving slowly and it is difficult to accurately measure plate motion.

- Turn on the **Earthquakes** layer.

Plate tectonics theory says that plates are rigid and will deform only at their boundaries. Earthquakes are indicators that a plate is deforming.

5. How well does plate tectonics theory predict the location of earthquakes? List three regions on Earth where locations of earthquakes do not occur on or near plate boundaries.

   a.

   b.

   c.

Identifying plate boundaries

Using earthquake data, magnetic patterns in the seafloor, and precise satellite navigation systems, researchers have collected information about how the plates are moving relative to one another.

- Turn off the **Earthquakes** layer.
- Turn on the **Relative Motion** layer.
- Select the **Relative Motion** layer.

The arrows in this layer are shown in pairs on opposite sides of the plate boundaries. They indicate the direction and speed of each plate relative to the other across the boundary. In each pair, the length of the arrows (the rate of plate motion) is the same, but their directions are opposite.

- Zoom in to an arrow of your choice and click the Identify tool.
- In the Identify Results window, select the **Relative Motion** layer from the list of layers.
- Using the Identify tool, to click on an arrow symbol to get precise rate information for that plate boundary. Close the Identify Results window when you are finished with it.

Using these arrows, you can determine if the two plates meeting at a boundary are colliding (convergent boundary), moving apart (divergent boundary), or slipping past each other (transform boundary). You will also need to recall what you've learned about the seismic, volcanic, and topographic patterns related to each type of boundary. These may help you identify plate boundaries where the arrows are absent or the plate motion or rate is uncertain.

Where to click?

Clicking on the arrow symbol can be tricky. When using the Identify tool on the **Relative Motion** layer's arrows, you need to click on the center of the arrow. Because the "arrowhead" is longer than the "shaft," the center is not where you expect it to be. The diagram below shows where to click.

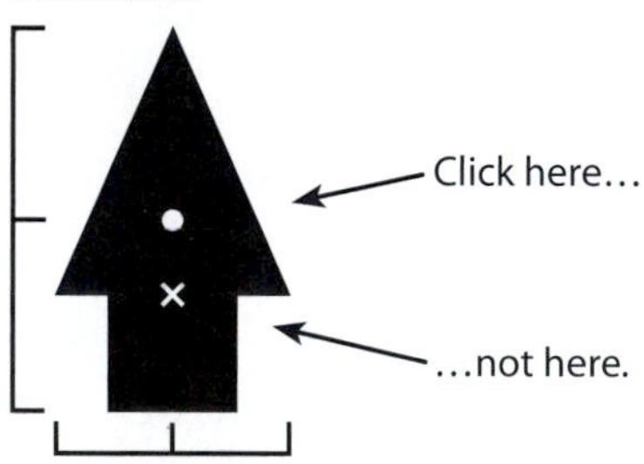

What is a rate?

A rate is a measure, or ratio, of a quantity per unit of something else. In this case, it is the measure of how fast the plates are moving (a length per unit time).

6. On the map (p. 28) you used for question 1, identify the type of each boundary as convergent, divergent, or transform using a symbol or color key. If you are not sure which type a particular boundary is, use your best guess.

7. Add the boundary type information to your map legend.

To see how well you predicted the types of boundaries:

- Click the QuickLoad button.
- Select **Data Frames**, choose **Boundary Types** from the list, and click **OK**.

This data frame shows the locations and types of plate boundaries on Earth.

- Select the **Plate Boundaries** layer.
- To see a diagram of each boundary type, click on a plate boundary using the Hyperlink tool.

8. Draw a box around each of the areas where your boundaries differ from the accepted map.

- Close the plate boundary diagram windows when finished.

9. How do you account for the differences between your predictions and the actual boundary types?

- Select a plate boundary of your choice within the data frame, and use the Zoom In tool to zoom in on it.

10. On your map (p. 28), draw a triangle around the plate boundary you have chosen.

11. Using all the data available in this activity, list and describe the evidence that indicates the type of boundary you have chosen. (Hint: Don't forget to include specific seismic, volcanic, and topographic clues.)

- Quit ArcMap and do not save changes.

Plate Boundaries

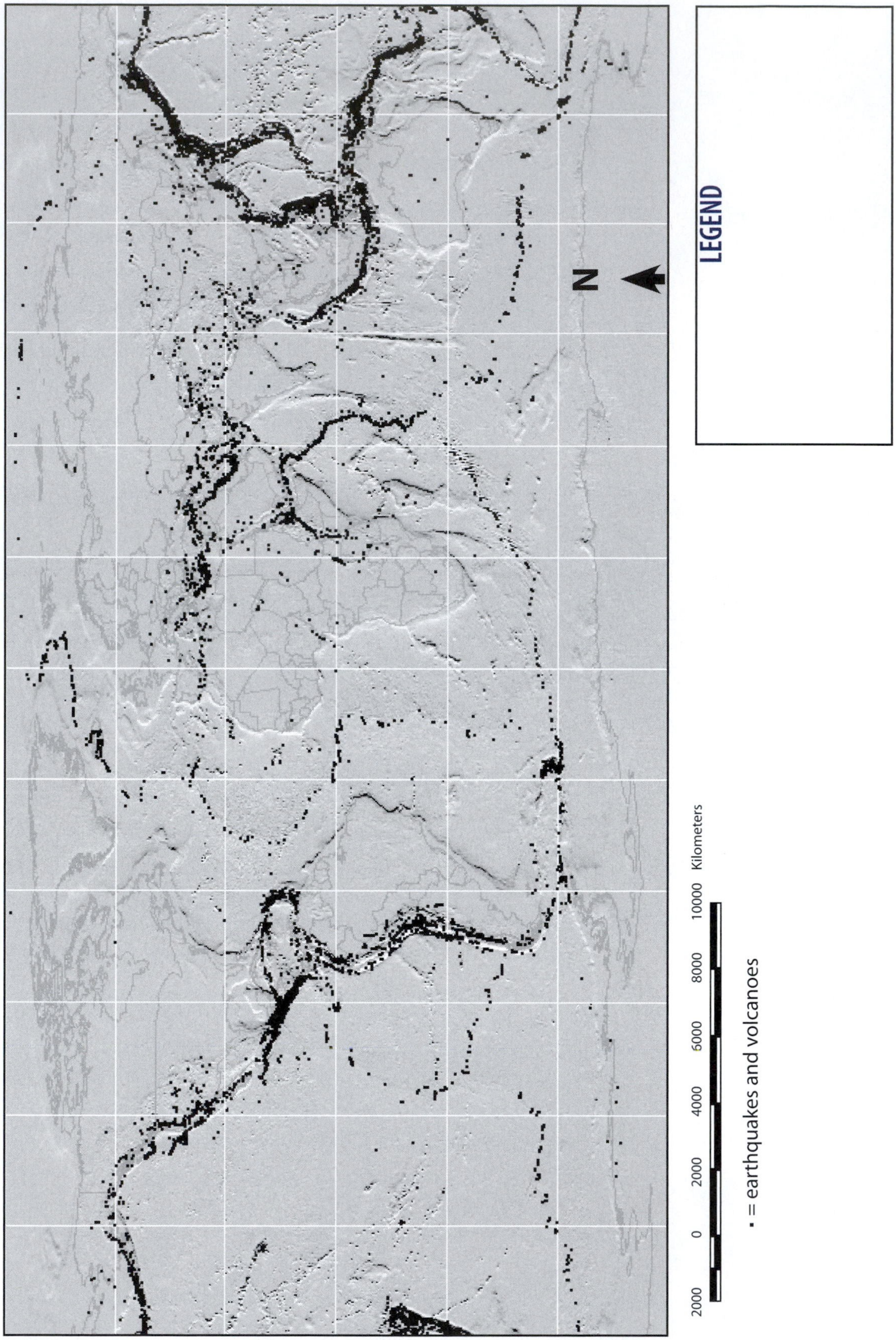

Unit 2
Exploring Plate Tectonics

In this unit, you will

- ***Calculate the rate of spreading of the Atlantic Ocean.***
- ***Investigate whether plate spreading rates change with time or vary across the globe.***
- ***Predict the ultimate fate of the Juan de Fuca plate.***
- ***Use the Hawaiian Islands to determine the rate of motion of the Pacific tectonic plate.***
- ***Determine when San Francisco and Los Angeles will finally be neighbors across the San Andreas Fault.***

Robert E. Wallace, USGS

Aerial view of the San Andreas fault slicing through the Carrizo Plain east of the city of San Luis Obispo, California.

Warm-up 2.1

Testing plate tectonics

You now know that Earth's plates are constantly growing, shrinking, and moving around. This means that Earth must have looked very different in the past.

Questions for thought and discussion

1. Earth's continents contain evidence that they were once in very different locations and may have had very different climates. List three signs or pieces of evidence that might indicate this.

   a.

   b.

   c.

Answer the following questions on your own or in a small discussion group, as assigned by your instructor.

2. What would you do to gather the evidence described above?

Thinking deeper (optional question)

Some scientists have suggested that even small changes in the shape, location, orientation, or topography of continents or oceans could significantly affect life on Earth.

3. Describe how even a small change in Earth's surface might have a major effect on life on the planet. Draw pictures to illustrate your point.

Investigation 2.2

Investigating seafloor age

In this activity, you will investigate evidence that tectonic plates change in size, shape, and location over time. This evidence suggests that plate tectonics has occurred for hundreds of millions of years.

Launch ArcMap, and locate and open the **etde_unit_2.mxd** file.

Refer to the tear-out Quick Reference Sheet located in the introduction to this module for GIS definitions and instructions on how to perform tasks.

In the Table of Contents, right-click the **Changing Plates** data frame and choose Activate.

Expand the **Changing Plates** data frame.

This data frame shows differences in the age of the rocks that form the ocean floor. These ages were determined using data from several sources, and have been summarized, mapped, and color-coded. The age bands generally run parallel to the spreading ridges. Seafloor age is a critical piece of evidence for plate tectonics. It shows us where new seafloor is created and how often. It can also be used to reconstruct how ocean basins have grown in the past, and how they may change in the future.

Spend a few minutes examining the seafloor age data. Use the **Seafloor Age** legend to answer these questions.

1. How many years does each colored band represent?

What does "m.y." stand for?

In the **Seafloor Age** legend, the abbreviation *m.y.* represents millons of years [ago] and is used to represent the age of material in past geologic time. For example, if a material is dated as "180 m.y.", it means it is 180 million years old.

2. What happens to the age of the seafloor in the Atlantic Ocean as distance increases from the Mid-Atlantic Ridge?

3. How long ago did the North Atlantic Ocean begin to "open up," or spread, near what is now the East Coast of the United States? Describe the reasoning you used to arrive at this answer.

4. Did the North Atlantic Ocean basin open up at the same time as the South Atlantic Ocean basin near South America? Describe evidence that supports your answer.

5. How could the widths of the age bands be used to figure out if the rate of plate spreading has been constant over time?

6. The oldest continental rocks formed 3.8 billion years ago. What is the age of the oldest seafloor?

Hint for question 7

Turn on the **Plate Boundaries** layer if necessary.

7. Why do you think there is such a difference in age between the oldest seafloor and the oldest continental rocks? (Hint: Near which type of plate boundary do you find the oldest oceanic rocks?)

8. During the time since the oldest continental rocks formed, about how many times could the world's ocean floors have been completely "recycled"? (Hint: Divide the age of the oldest continental rocks [3.8 billion years] by the age of the oldest seafloor [answer to question 6].)

Quit ArcMap and do not save changes.

Reading 2.3

Determining seafloor age

Paleomagnetism

Earth has a magnetic field that is like a bar magnet—one end is positive and one is negative. For reasons that are not fully understood, Earth's magnetic field changes **polarity**. That is, the north and south magnetic poles reverse. This process has occurred at uneven intervals, roughly once every 250,000 years over the past 5 million years.

As the ocean floor spreads at mid-ocean ridges, magma from the mantle wells up and solidifies onto the plate near the ridge. This process forms the volcanic rocks that make up the ocean floor.

When the newly formed **lithosphere** (made up of the solid upper mantle and Earth's crust) cools, some of the minerals become magnetized and align with Earth's magnetic field. This "fossil magnetism" preserves a record of the direction of Earth's magnetic field at the time the rocks formed.

Some bands of oceanic lithosphere formed when Earth's magnetic field was like it is today. We call this **normal polarity**. Other bands of oceanic lithosphere formed when Earth's magnetic field had the opposite polarity. We say these bands have **reversed polarity**.

The polarity changes of Earth's magnetic field are not regular. Instead they make complex sequences of normal and reversed polarity. The combination of seafloor spreading and changes in magnetic polarity form a unique pattern similar to the bar codes on products you buy at the store. Ships carrying sensitive equipment have crisscrossed the oceans recording the patterns of these fossil magnetic fields. Geologists can use the patterns to determine the age of the oceanic lithosphere.

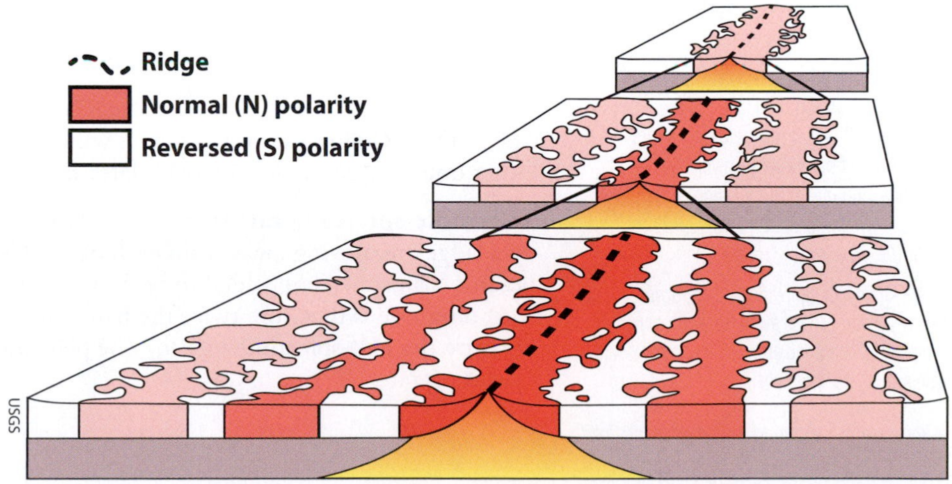

Constructing continents

Some plates remain relatively unchanged for millions of years, but most are growing or shrinking in size. Over Earth's 4.8-billion-year history, many oceanic plates have been destroyed and new ones have been created. Continental plates have been torn apart and smashed together into new arrangements, as in the Himalayas. Plate tectonics is reshaping the face of our planet.

Relative plate motion

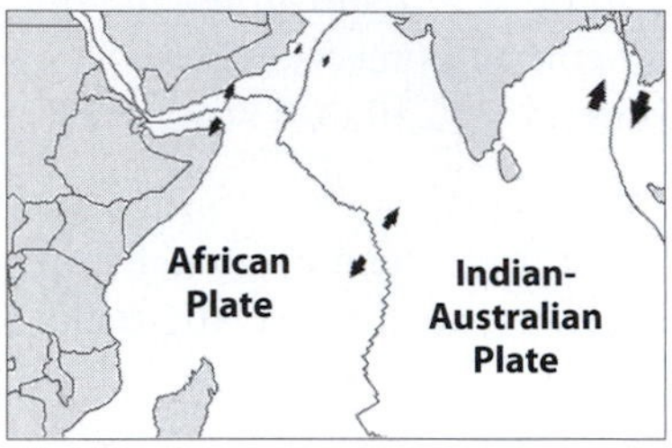

Arrows show the motion of the African plate and the Indian-Australian plate relative to each other.

Relative motion

All of Earth's plates are in motion. For this reason, it's often easier to describe a plate's motion relative to a neighboring plate.

For example, if you were standing on the eastern edge of the African plate looking at your dog Ralph (sitting very patiently) over on the bordering Indian-Australian plate, Ralph would appear to be moving away from you. To Ralph, *he* is the one sitting still while *you* move away from him. The motion is **relative to** the observer's point of view, or **frame of reference**. In actuality, however, this type of motion would occur over a very long period, geologically speaking.

Absolute plate motion

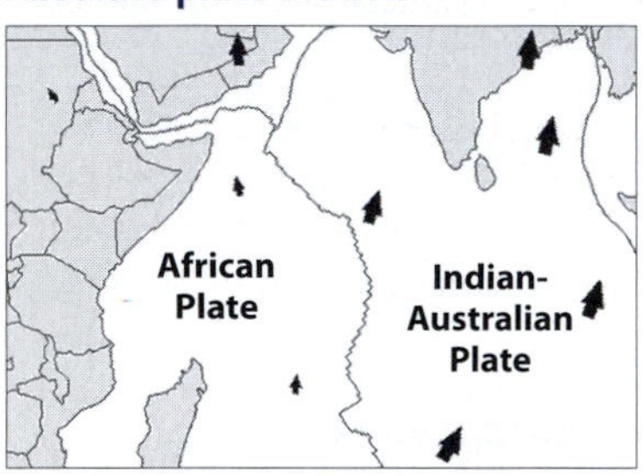

Arrows show the motion of the African plate and the Indian-Australian plate relative to a fixed reference point (a hot spot to the south).

Absolute motion and hot spots

When talking about entire plates, scientists often prefer to describe the plate motions relative to a common frame or point of reference. Ideally, this reference point is not moving like the plates but is **stationary**, or fixed, deep within Earth.

There are approximately 20 places around the Earth where the mantle below the lithosphere is unusually hot. These **hot spots** partially melt the plates above them, creating topographic bulges and chains of volcanoes. Although their origin is not completely understood, hot spots appear to be stationary and provide a frame of reference for measuring **absolute**, or true, plate motions.

Two of these hot spots are located in the United States. One is under the Big Island of Hawaii at the southeast end of the Hawaiian Island chain. The other is in Yellowstone National Park and is responsible for the volcanic and geothermal activity in that region.

Studying plate motions in a fixed frame of reference provides scientists with a better understanding of the forces that drive plate tectonics.

A Titanic example of spreading rates

The wreck of the ***Titanic*** rests at the bottom of the Atlantic Ocean to the west of the Mid-Atlantic Ridge. It moves about 15 mm farther west of that ridge each year. So the half-spreading rate at this location is 15 mm/yr. Thus, the whole-spreading rate is 2 x 15 mm or 30 mm/year.

As the Atlantic Ocean widens, the ***Titanic*** is slowly being carried away from where it originally left England at a rate of about 3 cm (a little over an inch) per year.

Spreading rates

Scientists measure spreading rates of oceanic lithosphere near spreading ridges as either a "half" or a "whole" rate.

The **half-spreading rate** is the rate at which newly formed lithosphere is added to either one or the other of the two plates separated by a spreading ocean ridge.

The **whole-spreading rate** is the combined rate at which the two plates separated by a ridge are moving away from each other. Usually, equal amounts of new oceanic lithosphere are added to both plates as they move apart. The whole-spreading rate is therefore twice the half-spreading rate. This means you can measure the half-spreading rate for one plate and double it to get the whole-spreading rate.

Changes in plate motion

Spreading rates away from spreading ridges and the directions of plate motions have changed many times during the past 200 million years. The Pacific plate experienced a particularly dramatic change in direction about 43 million years ago. If you examine the pattern of volcanoes extending northwest from Hawaii in the map below, you will notice that they form two intersecting chains, the Hawaiian and the Emperor volcanic chains.

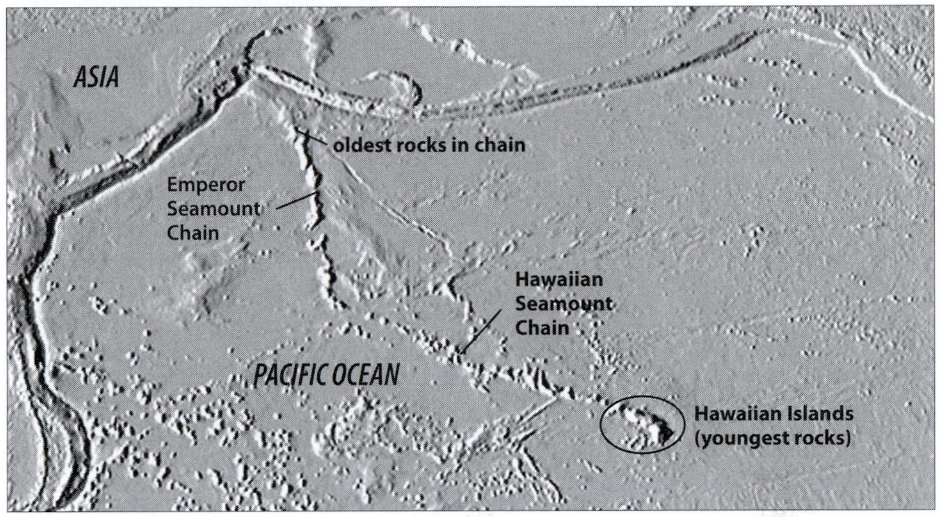

Seafloor topography of the northern Pacific Ocean basin.

By measuring the ages of rocks in each chain, we know that the "bend" in the chain occurred around 43 million years ago. Prior to that time, the Pacific plate was moving in a north-northwest direction, then abruptly (in geologic time, anyway) changed to a more west-northwest direction. Other seamount chains in the Pacific show a similar pattern.

Sudden changes in plate motion occur when two or more continents collide. Continental plates are too buoyant to be subducted. Instead, they plow into each other and pile up to form mountains. The Himalayas, the highest mountains on Earth, are being thrust upward as the Indian-Australian and the Eurasian plates collide. This has been going on continuously for the past 50 million years. Since the two plates first made contact, the rate of seafloor spreading in the Indian Ocean has decreased considerably.

Disappearing islands

The Hawaiian and Emperor volcanic chains mark the motion of the Pacific plate over a stationary hot spot. As each volcanic island is carried past the hot spot, wave action quickly erodes the exposed island completely away, leaving a submerged **seamount**.

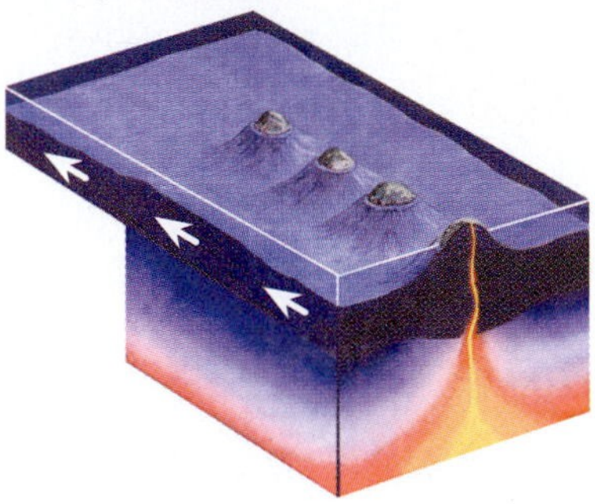

USGS

Earth's heat engine — slowly "running out of gas"

As Earth cools, the convection that drives the plates is gradually slowing down. Eventually — in billions of years — Earth will have lost most of its internal heat to space, and plate motions will slowly grind to a stop.

Questions

1. How are magnetic minerals in the oceanic lithosphere used to determine its age?

2. What is the difference between the relative motion and the absolute motion of a plate?

3. If the half-spreading rate at a plate boundary is 3 cm/yr, what would the whole-spreading rate be at that boundary?

4. What might cause a plate to slow its rate of movement or change direction?

Investigation 2.4

Investigating plate motion

In this section, you will measure the rate at which new lithosphere has formed along the Mid-Atlantic Ridge over the past 140 million years. Then, you will examine how this rate changed as the Atlantic Ocean opened.

Rate of spreading

To calculate the whole-spreading rate, you need to know the total width of new lithosphere formed and the time it took to create it.

m.y. — million years

$$\text{whole-spreading rate} = \frac{\text{total width of new lithosphere (km)}}{\text{time required to create new lithosphere (m.y.)}}$$

An easy way to do this is to measure the *half-spreading rate* — the width of new lithosphere created on one side of the spreading ridge — and double it.

Measuring new lithosphere

- Launch ArcMap, and locate and open the **etde_unit_2.mxd** file.

Refer to the tear-out Quick Reference Sheet located in the introduction to this module for GIS definitions and instructions on how to perform tasks.

- In the Table of Contents, right-click the **Changing Plates** data frame and choose Activate.
- Expand the **Changing Plates** data frame.

This data frame shows the spreading rates at spreading ridges located around the world. Each color band represents a 20-million-year time interval. To determine the width of new lithosphere created in the North Atlantic Ocean during each 20-million-year time interval:

- Turn on the **Latitude/Longitude**, **Plate Boundaries**, and **Seafloor Age** layers.
- Click the QuickLoad button.
- Select **Spatial Bookmarks**, choose **Atlantic Ocean Middle** from the list, and click **OK**.

The seafloor in the middle of the Atlantic Ocean should now be centered on your map.

- Using the Measure tool, click on one side of an interval, drag across to the other side of that interval, and double-click. For the 0 – 20-million-year interval (lightest), measure from the outer edge of the interval to the Mid-Atlantic Ridge (thick line, in the illustration on the next page). Measure perpendicularly to the spreading ridge, as shown in the illustration.
- Read the width (Total) of new lithosphere created during that time interval on the status bar, located at the bottom left corner of the map. (Your width will be different from the example below.)

Segment: 306.206774 Total: 306.206774 Kilometers

Measuring the first seafloor age interval

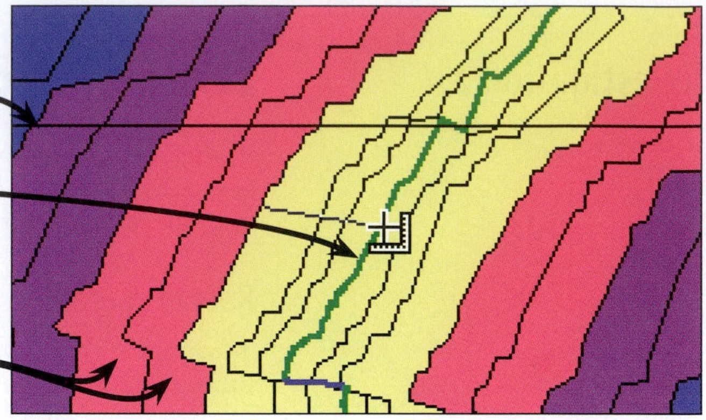

Locator map

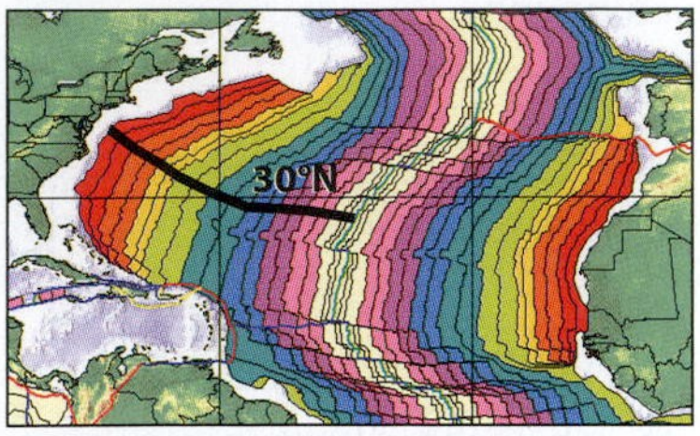

Path for measuring the spreading rate of the Mid-Atlantic Ridge in the North Atlantic Ocean.

1. Record your results in the **Width** column of the table provided below. Repeat this measurement for each of the other time intervals.

Computing the rates of spreading

Time before present *m.y.*	Width *km, measured*	Half-rate *mm/yr, calculated*	Whole-rate *mm/yr, calculated*
0-20			
21-40			
41-60			
61-80			
81-100			
101-120			
121-140			

Following the example shown at the left, perform the following calculations for each time interval listed in the left column of the table above.

2. Calculate the half-spreading rate, in mm/yr, and enter your answer in the **Half-rate** column. Convert units as necessary.

3. Calculate the whole-spreading rate, also in mm/yr, and enter your answer in the **Whole-rate** column of the table.

4. Use the bar graph on the next page to plot the **Half-rate** versus **Time before present** from the chart above.

5. How has the spreading rate changed over time, according to the bar graph?

Calculating spreading rates

Here's an example of calculating the spreading rates. Assuming the measured width is 260 km,

Width = 260 km

Half-Rate = Width/ Time

$= \frac{(260 \text{ ~~km~~ } \times 1{,}000{,}000 \text{ mm/~~km~~})}{20{,}000{,}000 \text{ yrs}}$

= 13 mm/yr

Whole-Rate = Half-Rate x 2

= 13 mm/yr x 2

= 26 mm/yr

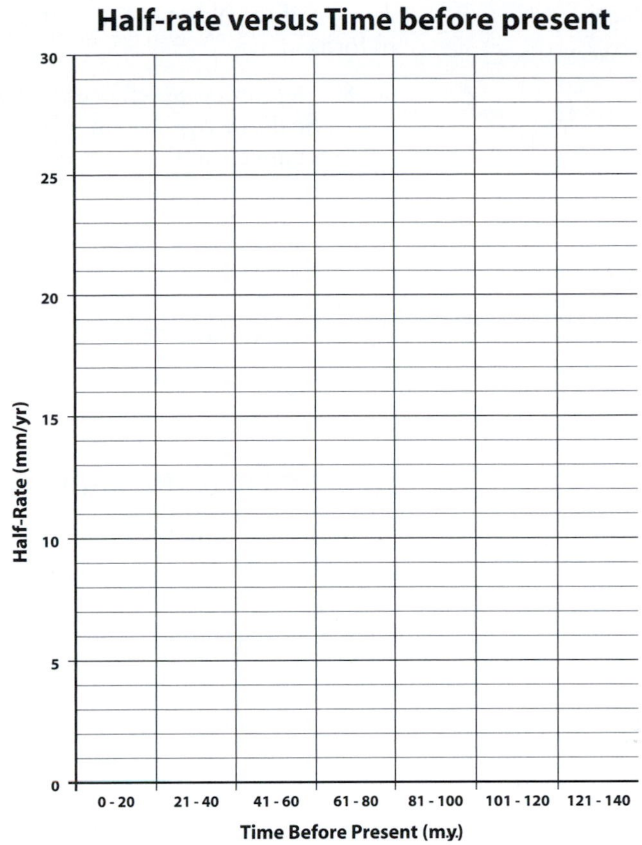

Compare the Atlantic to the Pacific

Are all spreading ridges alike? Next, you will compare the seafloor age patterns in the South Pacific and South Atlantic oceans.

You could calculate the rate of spreading in the South Pacific Ocean as you did for the North Atlantic Ocean, but you can also compare the rates by comparing the width of the age bands in the South Pacific to those in the North Atlantic. This data frame distorts distances at different latitudes, so compare age bands at similar latitudes (e.g., 30 degrees N and 30 degrees S).

Click the Full extent button to view the entire map.

6. Which ridge is spreading faster (Atlantic or Pacific)? How can you tell?

7. Examine the seafloor ages in the South Pacific. Was the spreading rate constant over time? Describe any changes in the spreading rate that have occurred over the past 120 million years.

Reading seafloor age

To read the age of the seafloor at any location, select the **Seafloor Age** layer and click on that location using the Identify tool. (Be sure to select the **Seafloor Age** layer in the Identify Results window.)

Which ocean is this?

If you aren't sure which ocean is which, turn on the **Ocean Labels** layer.

The age of the oldest rocks in an ocean basin indicates when the basin began to form.

8. Determine which ocean basin has the oldest of the old rocks, using the **Seafloor Age** legend. (This is the oldest ocean basin.) Which has the youngest of the old rocks? (This is the youngest ocean basin.)

   a. Oldest —

   b. Youngest —

Juan de Fuca's disappearing plate trick

The relative youth of the oldest oceanic lithosphere (less than 180 million years old) tells us that over the course of Earth's 4.8-billion-year history, many plates have come and gone. An example of a plate destined to disappear is the Juan de Fuca plate, located off the coast of Washington and Oregon. Assuming this plate continues moving at its current velocity, it should be possible to estimate when it will completely disappear.

Juan de Fuca plate

The Juan de Fuca plate is located off the western coast of North America.

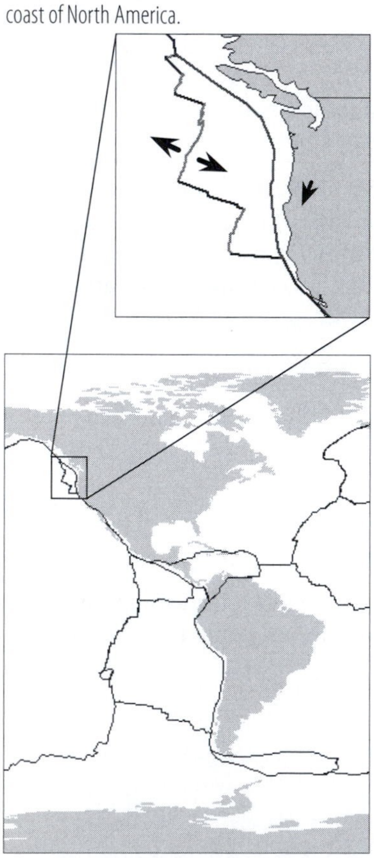

Next, you will zoom in on the Juan de Fuca plate, located just off the northwestern coast of the U.S. (see map at left):

- Click the QuickLoad button.
- Select **Spatial Bookmarks**, choose **Juan de Fuca Plate** from the list, and click **OK**.
- Turn on the **Plate Boundaries** layer.
- Turn on the **Plates** layer.

9. Label the types of plate boundaries on the map below.

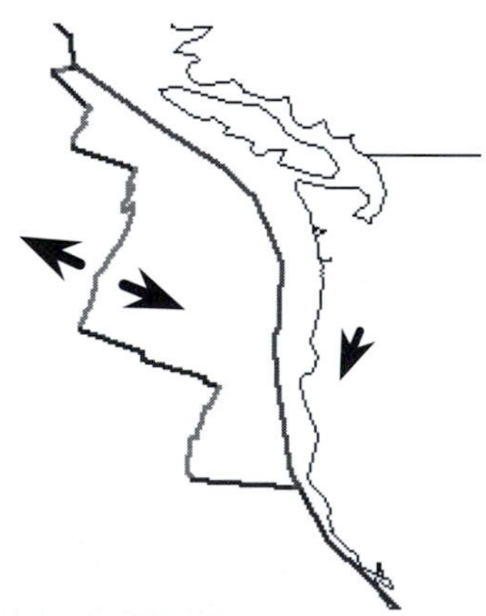

- Click the Identify tool.
- In the Identify Results window, select the **Plates** layer from the list of layers.
- Use the Identify tool to find the names of the three plates that form the plate boundaries.

10. On the map on the previous page, label the three plates with their names.

11. Under which plate is the Juan de Fuca plate being subducted?

Use the Measure tool to measure the widest part of the Juan de Fuca plate in a northwest-southeast direction, perpendicular to the spreading ridge.

12. Based on your measurement, how many kilometers of this plate must be subducted for the plate to disappear completely?

By measuring plate motions, scientists know that the new oceanic lithosphere is being formed at the spreading ridge on the western side of the Juan de Fuca plate at a rate of about 25 mm/year, while it is being subducted at the trench on its eastern side at a rate of about 30 mm/year.

13. Overall, how fast is this plate shrinking in an east-west direction each year?

Can you find the time?

Since you know that...

distance = speed × time

...you can rearrange the equation to find the time.

time = distance ÷ speed

14. At this rate, how long will it take for the Juan de Fuca plate to disappear completely?

a. Convert the rate you calculated in question 13 from mm/yr to km/yr.

b. Divide the distance (km) you calculated in question 12 by the rate (km/yr) you calculated in question 14a.

The Hawaiian-Emperor Volcanic Chain

You have seen that Earth's plates are in constant motion. How does this motion affect populated areas over time? In the next two sections, you will look at how plate motions are rearranging the United States. First you will examine the Hawaiian-Emperor Volcanic Chain to see what you can learn about plate motions.

Click the Full extent button to view the entire map.

Turn off all layers except **States**, **Countries**, and **Topography**.

Turn on the **Hawaiian Volcanoes (Wrap)** layer. Leave the **Hawaiian Volcanoes (No Wrap)** layer off for the moment.

The red symbols represent volcanoes in the Hawaiian-Emperor Volcanic Chain. Using radioactive dating of elements in Earth's crust, scientists have shown that the volcanoes were not all formed at the same time. Notice that the chain extends past the western (left) edge of the map and continues on the eastern (right) edge. Take a moment to explore the volcanoes.

- Turn off the **Hawaiian Volcanoes (Wrap)** layer.
- Turn on the **Hawaiian Volcanoes (No Wrap)** layer.
- Select the **Hawaiian Volcanoes (No Wrap)** layer.

In this layer, the volcanoes are grouped on one side of the map for your convenience.

- Click the QuickLoad button.
- Select **Spatial Bookmarks**, choose **Hawaii Volcanoes** from the list, and click **OK**.
- Click the Identify tool.
- In the Identify Results window, select the **Hawaiian Volcanoes (No Wrap)** layer from the list of layers.
- Use the Identify tool to find the age (in millions of years), latitude, and longitude of the volcano located at the northwestern-most (upper left) end of the chain. (Note: Negative latitude and longitude values indicate directions of south and west, respectively.)

15. Record the volcano's name, age, latitude, and longitude in the first row of the table below.

Name	Age *m.y.*	Latitude *°N – S*	Longitude *°E – W*

What is "S.M."?

The volcanoes that you identify in the **Hawaiian Volcanoes (No Wrap)** layer often have an "S.M." after their name. This simply refers to the term "seamount".

16. Record the names, ages, and geographic coordinates of five other volcanoes in the table. Choose volcanoes along the entire length of the chain. Make sure to go in order from northwest (upper left) to southeast (lower right), and to include the volcano at the southeastern-most end of the chain. (Zoom in, if needed, to identify this last volcano in the chain.)

17. What trend do you observe in the ages of the volcanoes as you follow the chain from the northwest to the southeast?

Assuming that the hot spot that formed these volcanoes is not moving, and that the motion of the Pacific plate over the hot spot is responsible for the existence of the chain, you can determine the rate at which the plate is moving.

- Use the Identify tool to locate Kinmei Seamount on the map. It's near the bend in the chain.

18. How long ago was Kinmei Seamount formed?

- Close the Identify Results window.
- Locate Kilauea Volcano at the southeast end of the island of Hawaii.
- Use the Measure tool to measure the distance between Kinmei Seamount and Kilauea.

19. How far did the Pacific plate move in the time between the formation of Kinmei Seamount and of Kilauea?

20. Assume the Pacific plate moved at a constant rate between the formation of Kinmei Seamount and Kilauea. At what rate has it moved? (See "Rate of Spreading" on page 39.)

- Turn on the **Absolute Motion** layer.
- Select the **Absolute Motion** layer.

This layer shows the speed and direction of plate motion relative to fixed locations (hot spots) for various points around the globe.

- Click the Identify tool.
- In the Identify Results window, select the **Absolute Motion** layer from the list of layers.
- Using the Identify tool, click on the brown arrow symbol just south of the Hawaiian islands. *(Important—see tip at left.)*
- Read the speed of the plate at that location in the Identify Results window.

21. How does the rate you calculated in the previous question compare to the absolute rate of motion near the Hawaiian Islands today?

- Close the Identify Results window.

Where to click?

Clicking on the arrow symbol can be tricky. When using the Identify tool on the **Absolute Motion** layer's arrows, you need to click on the center of the arrow. Because the "arrowhead" is longer than the "shaft," the center is not where you expect it to be. The diagram below shows where to click.

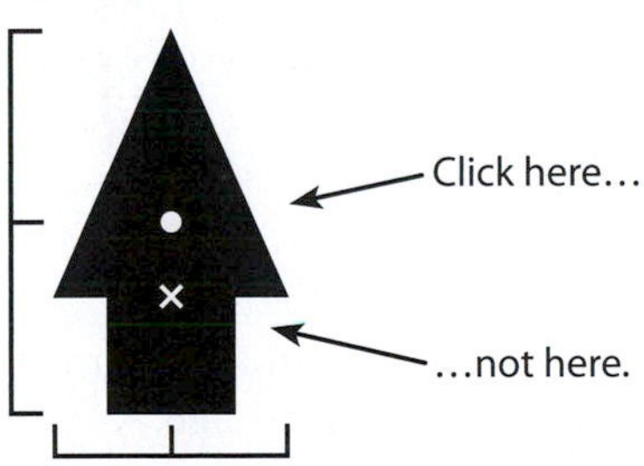

A (future) tale of two cities

Several of Earth's plate boundaries cut through heavily populated and economically important areas.

- Click the QuickLoad button.
- Select **Data Frames**, choose **Two Cities** from the list, and click **OK**.

The San Andreas Fault is a transform plate boundary that runs along the California coast. In time, the motion along this boundary will significantly change the face of the state and the fate of its two most famous cities.

- Click the Identify tool.

In the Identify Results window, select the **Plates** layer from the list of layers.

Use the Identify tool to find the names of the plates on which San Francisco and Los Angeles are located.

22. On which plates are the following cities located?

 a. San Francisco

 b. Los Angeles

Turn on the **Relative Motion** layer.

Select the **Relative Motion** layer.

Click the Identify tool.

In the Identify Results window, select the **Relative Motion** layer from the list of layers.

Use the Identify tool to find the relative speed of the Pacific plate. This is how fast Los Angeles is approaching San Francisco.

23. How fast (in mm/yr) is the Pacific plate moving northwest relative to the North American plate?

Where to click?

Clicking on the arrow symbol can be tricky. When using the Identify tool on the **Relative Motion** layer's arrows, you need to click on the center of the arrow. Because the "arrowhead" is longer than the "shaft," the center is not where you expect it to be. The diagram below shows where to click.

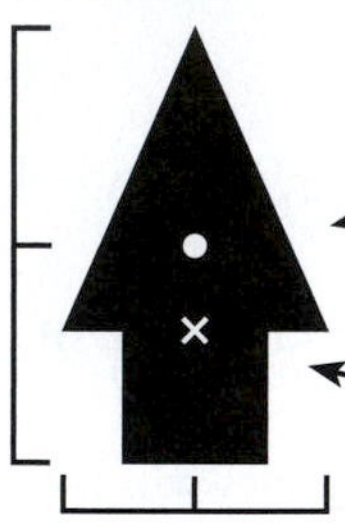

Use the Measure tool to measure the present distance (in meters) between San Francisco and Los Angeles as a straight line. Read the distance in the status bar.

24. How far apart are Los Angeles and San Francisco today? Record your answer in both meters and millimeters.

25. At their current rate of relative motion, how many years will it take San Francisco and Los Angeles to become suburbs of each other? (Hint: Use the equation at left. Don't forget to use the distance in millimeters from question 24.)

Can you find the time?

Since you know that...

$$\text{distance} = \text{speed} \times \text{time}$$

...you can rearrange the equation to find the time.

$$\text{time} = \text{distance} \div \text{speed}$$

26. There is a lot of evidence that tropical palm trees once grew in Antarctica and the southern parts of Chile and Argentina 30 million years ago. Explain how this could have happened when these places are so far from the tropics today.

Quit ArcMap and do not save changes.

Unit 3
Earthquake Hazards

In this unit, you will

- ***Examine earthquake data to locate the largest and most destructive earthquakes.***
- ***Investigate deadly earthquake patterns throughout history.***
- ***Discover which factors contribute most to an earthquake's destructive potential.***
- ***Explore the relationship between population, national wealth, and seismic risk.***

C. J. Langer, USGS

On December 7, 1988 a magnitude 6.9 earthquake struck northwestern Armenia, killing as many as 50,000 people and leaving over 500,000 homeless. Many historical buildings, such as this church in Leninakan, either collapsed or sustained severe damage. High, unsupported roofs make churches particularly vulnerable to earthquake damage.

Warm-up 3.1

Location map

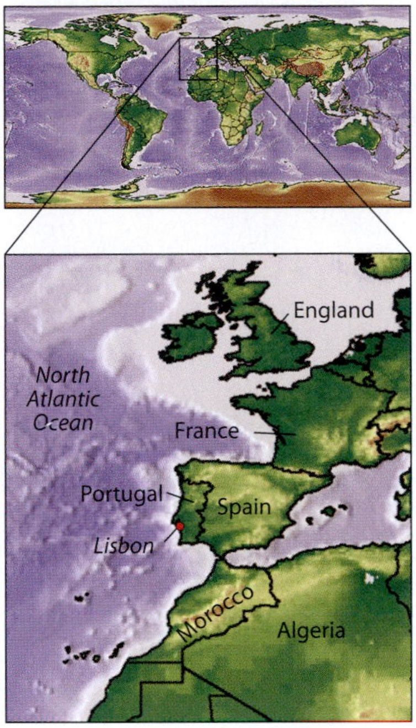

The Great Lisbon Earthquake

After reading the story of the earthquake that struck Lisbon, Portugal in 1755 (page 53), list and describe all of the earthquake-related hazards discussed. Feel free to add other hazards from your previous knowledge or experience with earthquakes.

Earthquake hazards

1.

2.

3.

4.

5.

6.

7.

8.

Earthquake hazard examples (optional)

If you have access to a computer, you can see examples of these and other earthquake hazards. Be sure to add any new hazards you find to your list.

- Launch ArcMap, and locate and open the **etde_unit_3.mxd** file.

Refer to the tear-out Quick Reference Sheet located in the introduction to this module for GIS definitions and instructions on how to perform tasks.

- In the Table of Contents, right-click the **Geological Hazards** data frame and choose Activate.
- Expand the **Geological Hazards** data frame.

To see examples of other earthquake hazards:

- Turn on the **Hazard Links** layer.

- Select the **Hazard Links** layer.
- Using the Hyperlink tool, click on each of the orange earthquake hazard symbols (leaning buildings) on the map.
- Read the caption for each picture, then close its window. There may be more than one picture for each link.

Questions

1. Which of the hazards you listed caused the greatest amount of damage and cost the most lives in the Lisbon earthquake?

2. What do you think could have been done to reduce the loss of life and property in Lisbon? Explain.

Before 1755, Portugal was a major world power, controlling a vast global empire. In a weakened state after the earthquake, Portugal lost wars to Spain and England. Today, Portugal is the poorest country in Western Europe.

3. How might the world be different today if this earthquake had occurred somewhere else, or had been less severe?

4. If an earthquake of this magnitude occurred in a major U.S. city today, how might it affect you?

Eyewitness account by Rev. Charles Davy

The Earthquake at Lisbon, 1755

That was the year when Lisbon-town / Saw the earth open and gulp her down ("The Wonderful One-Hoss Shay," by Oliver Wendell Holmes)

THERE never was a finer morning seen than the 1st of November; the sun shone out in its full luster; the whole face of the sky was perfectly serene and clear; and not the least signal of warning of that approaching event, which has made this once flourishing, opulent, and populous city a scene of the utmost horror and desolation, except only such as served to alarm, but scarcely left a moment's time to fly from the general destruction.

It was on the morning of this fatal day, between the hours of nine and ten, that I was set down in my apartment, just finishing a letter, when the papers and table I was writing on began to tremble with a gentle motion, which rather surprised me, as I could not perceive a breath of wind stirring. Whilst I was reflecting with myself what this could be owing to, but without having the least apprehension of the real cause, the whole house began to shake from the very foundation, which at first I imputed to the rattling of several coaches in the main street, which usually passed that way, at this time, from Belem to the palace; but on hearkening more attentively, I was soon undeceived, as I found it was owing to a strange frightful kind of noise under ground, resembling the hollow distant rumbling of thunder. All this passed in less than a minute, and I must confess I now began to be alarmed, as it naturally occurred to me that this noise might possibly be the forerunner of an earthquake, as one I remembered, which had happened about six or seven years ago, in the island of Madeira, commenced in the same manner, though it did little or no damage.

Upon this I threw down my pen and started upon my feet, remaining a moment in suspense, whether I should stay in the apartment or run into the street, as the danger in both places seemed equal; and still flattering myself that this tremor might produce no other effects than such inconsiderable ones as had been felt at Madeira; but in a moment I was roused from my dream, being instantly stunned with a most horrid crash, as if every edifice in the city had tumbled down at once. The house I was in shook with such violence that the upper stories immediately fell; and though my apartment (which was the first floor) did not then share the same fate, yet everything was thrown out of its place in such a manner that it was with no small difficulty I kept my feet, and expected nothing less than to be soon crushed to death, as the walls continued rocking to and fro in the frightfulest manner, opening in several places; large stones falling down on every side from the cracks, and the ends of most of the rafters starting out from the roof. To add to this terrifying scene, the sky in a moment became so gloomy that I could now distinguish no particular object; it was an Egyptian darkness indeed, such as might be felt; owing, no doubt, to the prodigious clouds of dust and lime raised from so violent a concussion, and, as some reported, to sulphureous exhalations, but this I cannot affirm; however, it is certain I found myself almost choked for near ten minutes.

From *The Volcano's Deadly Work*, C. Morris.

I hastened out of the house and through the narrow streets, where the buildings either were down or were continually falling, and climbed over the ruins of St. Paul's Church to get to the river's side, where I thought I might find safety. Here I found a prodigious concourse of people of both sexes, and of all ranks and conditions, among whom I observed some of the principal canons of the patriarchal church, in their purple robes and rochets, as these all go in the habit of bishops; several priests who had run from the altars in their sacerdotal vestments in the midst of their celebrating Mass; ladies half dressed, and some without shoes; all these, whom their mutual dangers had here assembled as to a place of safety, were on their knees at prayers, with the terrors of death in their countenances, every one striking his breast and crying out incessantly, Miserecordia meu Dios! . . . In the midst of our devotions, the second great shock came on, little less violent than the first, and completed the ruin of those buildings which had been already much shattered. The consternation now became so universal that the shrieks and cries of Miserecordia could be distinctly heard from the top of St. Catherine's Hill, at a considerable distance off, whither a vast number of people had likewise retreated; at the same time we could hear the fall of the parish church there, whereby many persons were killed on the spot, and others mortally wounded.

You may judge of the force of this shock, when I inform you it was so violent that I could scarce keep on my knees; but it was attended with some circumstances still more dreadful than the former. On a sudden I heard a general outcry, "The sea is coming in, we shall be all lost." Upon this, turning my eyes towards the river, which in that place is nearly four miles broad, I could perceive it heaving and swelling in the most unaccountable manner, as no wind was stirring. In an instant there appeared, at some small distance, a large body of water, rising as it were like a mountain. It came on foaming and roaring, and rushed towards the shore with such impetuosity, that we all immediately ran for our lives as fast as possible; many were actually swept away, and the rest above their waist in water at a good distance from the banks. For my own part I had the narrowest escape, and should certainly have been lost, had I not grasped a large beam that lay on the ground, till the water returned to its channel, which it did almost at the same instant, with equal rapidity. As there now appeared at least as much danger from the sea as the land, and I scarce knew whither to retire for shelter, I took a sudden resolution of returning back, with my clothes all dripping, to the area of St. Paul's. Here I stood some time, and observed the ships tumbling and tossing about as in a violent storm; some had broken their cables, and were carried to the other side of the Tagus; others were whirled around with incredible swiftness; several large boats were turned keel upwards; and all this without any wind, which seemed the more astonishing. It was at the time of which I am now speaking, that the fine new quay [*a wharf or reinforced shore where ships were loaded and unloaded*], built entirely of rough marble, at an immense expense, was entirely swallowed up, with all the people on it, who had fled thither for safety, and had reason to think themselves out of danger in such a place: at the same time, a great number of boats and small vessels, anchored near it (all likewise full of people, who had retired thither for the same purpose), were all swallowed up, as in a whirlpool, and nevermore appeared.

This last dreadful incident I did not see with my own eyes, as it passed three or four stones' throws from the spot where I then was; but I had the account as here given from several masters of ships, who were anchored within two or three hundred yards of the quay, and saw the whole catastrophe. One of them in particular informed me that when the second shock came on, he could perceive the whole city waving backwards and forwards, like the sea when the wind first begins to rise; that the agitation of the earth was so great even under the river, that it threw up his large anchor from the mooring, which swam, as he termed it, on the surface of the water: that immediately upon this extraordinary concussion, the river rose at once near twenty feet, and in a moment subsided; at which instant he saw the quay, with the whole concourse of people upon it, sink down, and at the same time every one of the boats and vessels that were near it was drawn into the cavity, which he supposed instantly closed upon them, inasmuch as not the least sign of a wreck was ever seen afterwards. This account you may give full credit to, for as to the loss of the vessels, it is confirmed by everybody; and with regard to the quay, I went myself a few days after to convince myself of the truth, and could not find even the ruins of a place where I had taken so many agreeable walks, as this was the common rendezvous of the factory in the cool of the evening. I found it all deep water, and in some parts scarcely to be fathomed.

This is the only place I could learn which was swallowed up in or about Lisbon, though I saw many large cracks and fissures in different parts; and one odd phenomenon I must not omit, which was communicated to me by a friend who has a house and wine cellars on the other side of the river, viz., that the dwelling-house being first terribly shaken, which made all the family run out, there presently fell down a vast high rock near it; that upon this the river rose and subsided in the manner already mentioned, and immediately a great number of small fissures appeared in several contiguous pieces of ground, from whence there spouted out, like a jet stream, a large quantity of fine white sand to a prodigious height. It is not to be doubted the bowels of the earth must have been excessively agitated to cause these surprising effects; but whether the shocks were owing to any sudden explosion of various minerals mixing together, or to air pent up, and struggling for vent, or to a collection of subterranean waters forcing a passage, God only knows. As to the fiery eruptions then talked of, I believe they are without foundation, though it is certain I heard several complaining of strong sulphureous smells, a dizziness in their heads, a sickness in their stomachs, and difficulty of respiration, not that I felt any such symptoms myself.

I had not been long in the area of St. Paul's when I felt the third shock, somewhat less violent than the two former, after which the sea rushed in again, and retired with the same rapidity, and I remained up to my knees in water, though I had gotten upon a small eminence at some distance from the river, with the ruins of several intervening houses to break its force. At this time I took notice the waters retired so impetuously, that some vessels were left quite dry, which rode in seven fathom water; the river thus continued alternately rushing on and retiring several times together, in such sort that it was justly dreaded Lisbon would now meet the same fate which a few years before had befallen the city of Lima; and no doubt had this place lain open to the sea, and the force of the waves not been somewhat broken by the winding of the bay, the lower parts of it at least would have been totally destroyed.

The master of a vessel which arrived here just after the 1st of November, assured me that he really concluded he had struck upon a rock, till he threw out the lead, and could find no bottom, nor could he possibly guess at the cause, till the melancholy sight of this desolate city left him no room to doubt of it. The two first shocks, in fine, were so violent that several pilots were of opinion the situation of the bar at the mouth of the Tagus was changed. Certain it is that one vessel, attempting to pass through the usual channel, foundered, and another struck on the sands, and was at first given over for lost, but at length got through. There was another great shock after this, which pretty much affected the river, but I think not so violently as the preceding; though several persons assured me that as they were riding on horseback in the great road leading to Belem, one side of which lies open to the river, the waves rushed in with so much rapidity that they were obliged to gallop as fast as possible to the upper grounds, for fear of being carried away.

I was now in such a situation that I knew not which way to turn myself: if I remained there, I was in danger from the sea; if I retired farther from the shore, the houses threatened certain destruction; and at last, I resolved to go to the

Mint, which being a low and very strong building, had received no considerable damage, except in some of the apartments towards the river. The party of soldiers, which is every day set there on guard, had all deserted the place, and the only person that remained was the commanding officer, a nobleman's son, of about seventeen or eighteen years of age, whom I found standing at the gate. As there was still a continued tremor of the earth, and the place where we now stood (being within twenty or thirty feet of the opposite houses, which were all tottering) appeared too dangerous, the courtyard being likewise full of water, we both retired inward to a hillock of stones and rubbish: here I entered into conversation with him, and having expressed my admiration that one so young should have the courage to keep his post, when every one of his soldiers had deserted theirs, the answer he made was, though he were sure the earth would open and swallow him up, he scorned to think of flying from his post. In short, it was owing to the magnanimity of this young man that the Mint, which at this time had upwards of two millions of money in it, was not robbed; and indeed I do him no more than justice in saying that I never saw any one behave with equal serenity and composure on occasions much less dreadful than the present.

Perhaps you may think the present doleful subject here concluded; but alas! the horrors of the 1st of November are sufficient to fill a volume. As soon as it grew dark, another scene presented itself little less shocking than those already described: the whole city appeared in a blaze, which was so bright that I could easily see to read by it. It may be said without exaggeration, it was on fire at least in a hundred different places at once, and thus continued burning for six days together, without intermission, or the least attempt being made to stop its progress.

It went on consuming everything the earthquake had spared, and the people were so dejected and terrified that few or none had courage enough to venture down to save any part of their substance; every one had his eyes turned towards the flames, and stood looking on with silent grief, which was only interrupted by the cries and shrieks of women and children calling on the saints and angels for succor, whenever the earth began to tremble, which was so often this night, and indeed I may say ever since, that the tremors, more or less, did not cease for a quarter of an hour together. I could never learn that this terrible fire was owing to any subterranean eruption, as some reported, but to three causes, which all concurring at the same time, will naturally account for the prodigious havoc it made. The 1st of November being All Saints' Day, a high festival among the Portuguese, every altar in every church and chapel (some of which have more than twenty) was illuminated with a number of wax tapers and lamps as customary; these setting fire to the curtains and timber-work that fell with the shock, the conflagration soon spread to the neighboring houses, and being there joined with the fires in the kitchen chimneys, increased to such a degree that it might easily have destroyed the whole city though no other cause had concurred, especially as it met with no interruption.

But what would appear incredible to you, were the fact less public and notorious, is that a gang of hardened villains, who had been confined and got out of prison when the wall fell, at the first shock, were busily employed in setting fire to those buildings which stood some chance of escaping the general destruction. I cannot conceive what could have induced them to this hellish work, except to add to the horror and confusion that they might, by this means, have the better opportunity of plundering with security. But there was no necessity for taking this trouble, as they might certainly have done their business without it, since the whole city was so deserted before night that I believe not a soul remained in it except those execrable villains and others of the same stamp. It is possible some among them might have had other motives besides robbing, as one in particular being apprehended (they say he was a Moor, condemned to the galleys), confessed at the gallows, that he had set fire to the king's palace with his own hand; at the same time glorying in the action, and declaring with his last breath that he hoped to have burnt all the royal family. It is likewise generally believed that Mr. Bristow's house, which was an exceedingly strong edifice, built on vast stone arches, and had stood the shocks without any great damage further than what I have mentioned, was consumed in the same manner. The fire, in short, by some means or other, may be said to have destroyed the whole city, at least everything that was grand or valuable in it.

With regard to the buildings, it was observed that the solidest in general fell the first. Every parish church, convent, nunnery, palace, and public edifice, with an infinite number of private houses, were either thrown down or so miserably shattered that it was rendered dangerous to pass by them.

The whole number of persons that perished, including those who were burnt or afterwards crushed to death whilst digging in the ruins, is supposed, on the lowest calculation, to amount to more than sixty thousand; and though the damage in other respects cannot be computed, yet you may form some idea of it when I assure you that this extensive and opulent city is now nothing but a vast heap of ruins; that the rich and the poor are at present upon a level; some thousands of families which but the day before had been easy in their circumstances, being now scattered about in the fields, wanting every conveniency of life, and finding none able to relieve them.

Source—Tappan, E. M. (Ed). (1914). The World's Story: A History of the World in Story, Song, and Art: Vol V. Italy, France, Spain, and Portugal (pp. 618-628). Boston: Houghton Mifflin.

Investigation 3.2

Deadly earthquakes

Understanding where and when a major earthquake may strike is critical to assessing risk to life and property. It is also useful in developing plans to reduce that risk.

In this activity, you will investigate questions like "Where do the most damaging earthquakes occur?" You will also compare the locations of large earthquakes with the locations of deadly earthquakes and investigate changes in the distribution of deadly earthquakes over time.

Launch ArcMap, and locate and open the **etde_unit_3.mxd** file.

Refer to the tear-out Quick Reference Sheet located in the introduction to this module for GIS definitions and instructions on how to perform tasks.

In the Table of Contents, right-click the **Earthquake Hazards** data frame and choose Activate.

Expand the **Earthquake Hazards** data frame.

This map shows the locations of nearly 12,000 earthquakes of magnitude 5 and greater. They are a representative sample of over 300,000 quakes recorded around the world since 1973.

Large earthquakes

Most damage occurs from earthquakes with magnitudes of 6.5 and greater.

Click the Select By Attributes button.

To see the locations of these powerful earthquakes (of magnitude 6.5 or greater), query the **Earthquakes** layer for **("Magnitude" >= 6.5)** as shown in steps 1–6:

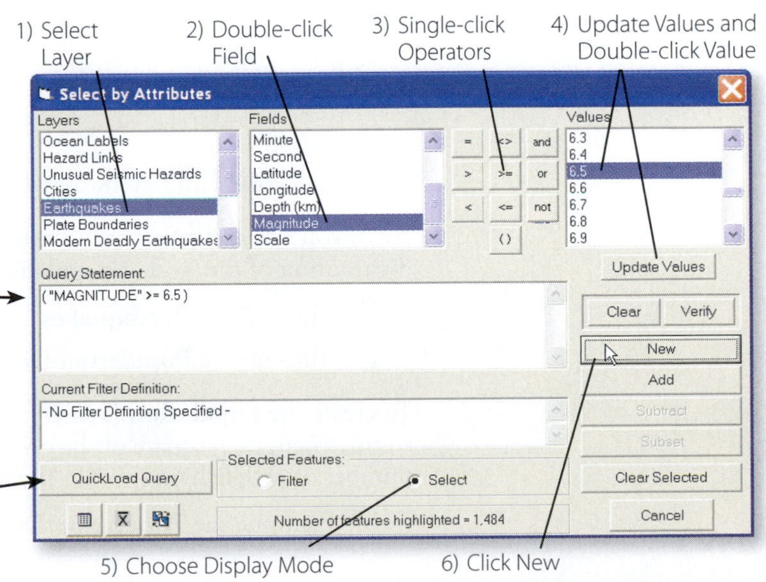

Read query statement here as you enter it.

QuickLoad Query

- Click the QuickLoad Query button and select the **Earthquake Hazards - Large Earthquakes (>=mag 6.5)** query.
- Click **OK**.
- Click **New**.

- If you have difficulty entering the query statement correctly, refer to the **QuickLoad Query** described at left on the previous page.
- Close the Select By Attributes window.

Now the strong earthquakes are highlighted on your map.

- Turn on the **Plate Boundaries** layer.
- Turn the **Earthquakes** layer on and off to examine the relationship between the large earthquakes and the plate boundaries.

1. At which type of boundary are large-magnitude earthquakes most common?

- Click the Clear Selected Features button.

Deadly earthquakes

- Turn off the **Plate Boundaries** and **Earthquakes** layers.
- Turn on the **Deadly Earthquakes (Deaths)** layer.
- Select the **Deadly Earthquakes (Deaths)** layer.

The **Deadly Earthquakes (Deaths)** layer shows the locations of historical earthquakes in which one or more people died. Many of these events were magnitude 6 or greater, but quakes as small as magnitude 4 caused fatalities and significant damage. Prior to the early 1900s, instruments used to measure earthquakes did not record magnitudes, which were estimated much later from written reports of earthquake damage.

- Turn the **Deadly Earthquakes (Deaths)** and **Earthquakes** layers on and off, and zoom in and out as needed to answer the following question.

2. How do the locations of the deadly earthquakes compare to the locations of the large earthquakes?

- Click the Full Extent button to see the entire map.

Next, you will compare the locations of deadly earthquakes to the current distribution of the world's population.

- Turn off the **Earthquakes** and **Topography** layers.
- Turn on the **Population Density** and **Deadly Earthquake** layers.

To create the **Population Density** layer, Earth's land area was first divided into a grid. Then, each square of the grid was filled with a color that represents the number of people living within the square.

- Examine the **Population Density** legend.

3. Which color represents high population density? Which color represents low population density?

   a. High density—

   b. Low density—

4. In general, how does the population density compare to the locations of deadly earthquakes?

5. Aside from earthquake magnitude and population density, what factors do you think might contribute to a higher damage and death toll in a large earthquake?

- Turn off the **Population Density** layer.
- Turn on the **Topography** layer.

Deadly earthquakes throughout history

Next you will look at how the pattern of deadly earthquakes has changed over the past two centuries. You will determine the number and locations of deadly earthquakes around the world during four time periods from 186 B.C. to the present.

- Click the Select By Attributes button.
- To display only the deadly earthquakes that occurred before A.D. 500, query the **Deadly Earthquakes (Deaths)** layer using the following query statement (you will need to type the value "500"):

  ("YEAR" < 500)

- Click the **New** button.
- If you have difficulty entering the query statement correctly, refer to the **QuickLoad Query** described at left.

Now the deadly earthquakes that occurred before A.D. 500 (between 186 B.C. and A.D. 499) are highlighted on your map.

- Click the Statistics button in the Select By Attributes window.
- In the Statistics window, calculate statistics for **only selected features** of the **Deadly Earthquakes (Deaths)** layer, using the **Deaths** field.

What are "B.C." and "A.D."?

B.C. stands for Before Christ and refers to the time period before Christ's birth. ***A.D.*** stands for Anno Domini (Latin for "in the year of the Lord") and refers to the time period after Christ's birth.

QuickLoad Query

- Click the QuickLoad Query button and select the **Deadly Earthquakes 1 - Year < 500 AD** query.
- Click **OK**.
- Click **New**.

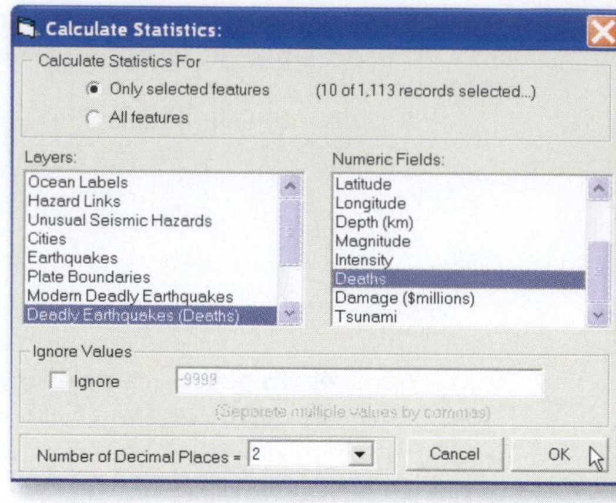

- Click **OK**. Be patient while the statistics are calculated.

6. Record the number of quakes (**Number of Records**), number of deaths (**Total**), and deaths per quake (**Mean**) in the table on the following page. In the last row of the table, identify the region(s) of the world where the earthquakes occurred.

- Close the Statistics window.
- Click the **Clear** button in the Select By Attributes window.
- To display only the deadly earthquakes that occurred between A.D. 500 and A.D. 999, query the **Deadly Earthquakes (Deaths)** layer using the following query statement:

("YEAR" >= 500) AND ("YEAR" < 1000)

- Click the **New** button.
- If you have difficulty entering the query statement correctly, refer to the **QuickLoad Query** described at left.

QuickLoad Query

- Click the QuickLoad Query button and select the **Deadly Earthquakes 2 - Year 500 – 900 AD** query.
- Click **OK**.
- Click **New**.

Now the deadly earthquakes that occurred between A.D. 500 and 999 are highlighted on your map.

7. Repeat the statistics procedure above and record the number of quakes, deaths, and deaths per quake in the table on the following page. Identify the region(s) where the earthquakes occurred.

- Close the Statistics window when you are finished.
- Click the **Clear** button in the Select By Attributes window.
- To display only the deadly earthquakes that occurred between A.D. 1000 and 1499, query the **Deadly Earthquakes (Deaths)** layer using the following query statement:

("YEAR" >= 1000) AND ("YEAR" < 1500)

- Click the **New** button.
- If you have difficulty entering the query statement correctly, refer to the **QuickLoad Query** described at left.

QuickLoad Query

- Click the QuickLoad Query button and select the **Deadly Earthquakes 3 - Year 1000 – 1499 AD** query.
- Click **OK**.
- Click **New**.

Now the deadly earthquakes that occurred between A.D. 1000 and 1499 are highlighted on your map.

8. Repeat the statistics procedure and record the number of quakes, deaths, and deaths per quake in the table below. Identify the region(s) where the earthquakes occurred.

Close the Statistics window when you are finished.

Click the **Clear** button in the Select By Attributes window.

To display only the deadly earthquakes that occurred between A.D. 1500 and 2005, query the **Deadly Earthquakes (Deaths)** layer using the following query statement:

("YEAR" >= 1500)

Click the **New** button.

If you have difficulty entering the query statement correctly, refer to the **QuickLoad Query** described at left.

QuickLoad Query

- Click the QuickLoad Query button and select the **Deadly Earthquakes 4 - Year 1500 to Present** query.
- Click **OK**.
- Click **New**.

Now the deadly earthquakes that occurred between A.D. 1500 and 2005 are highlighted on your map.

9. Repeat the statistics procedure and record the number of quakes, deaths, and deaths per quake in the following table. Identify the region(s) where the earthquakes occurred.

Time period	186 B.C. – A.D. 499	A.D. 500 – 999	A.D. 1000 – 1499	A.D. 1500 – 2005
Number of years	686	500	500	505
Number of quakes *(Statistics - Number of Records)*				
Number of deaths *(Statistics - Total)*				
Deaths per quake *(Statistics - Mean)*				
Deaths per year *(Calculate - see example)*				
Region(s) *(Identify from map)*				

Close the Statistics and Select By Attributes windows when you are finished.

10. Calculate the average number of deaths per year for each period and record it in the table. (Hint: See example calculation at left. Use the space below for your calculations.)

Calculating the average number of deaths per year

Average Deaths per year =

Total Number of deaths ÷ number of years

Example

467,960 deaths ÷ 686 years

average = 682 deaths/year

Click the Clear Selected Features button.

11. Describe how the locations of the deadly earthquakes have changed over time.

12. How have the numbers of deadly earthquakes changed over time, according to the table?

13. How would the ability to record data throughout history affect the pattern of deadly earthquakes shown here? Give an example.

14. Plot the average number of deaths per year (from the table on page 63) for each time period as a bar graph in the graph below.

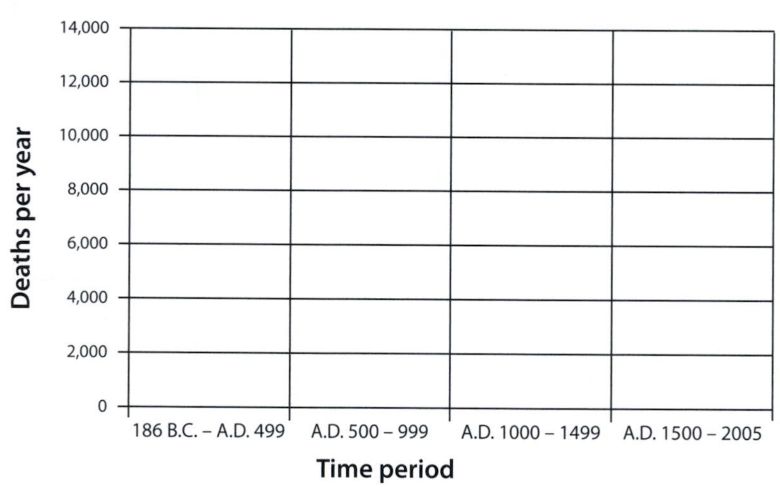

15. Plot the average number of deaths per earthquake (from the table on page 63) for each time period as a bar graph in the graph below.

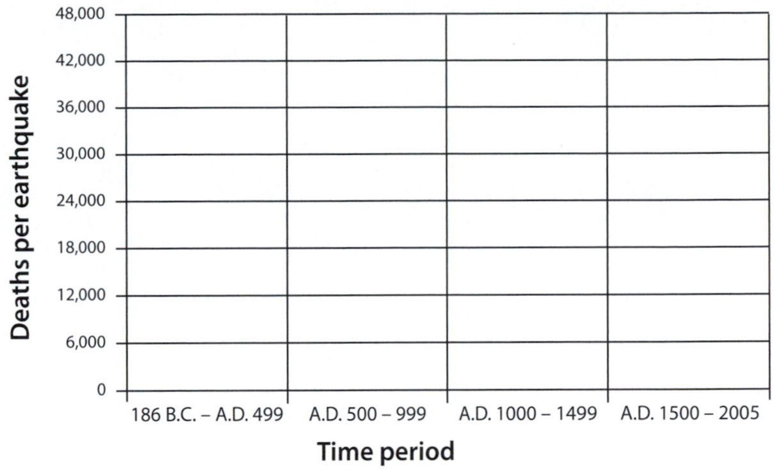

16. How have the average numbers of deaths per year and deaths per earthquake changed over time?

17. How might changes in the population distribution and improvements in earthquake detection contribute to these patterns?

18. Do these data show that the number of strong earthquakes has been increasing over the past 2000 years? Explain.

Quit ArcMap and do not save changes.

Reading 3.3

Seismic hazards

Hazards versus risks

The words **hazard** and **risk** are often used interchangeably, but they mean distinctly different things. A hazard is a potentially dangerous event or process. Risk is the potential loss of life, property, or production capacity due to a hazard. Geological hazards exist even if lives and property are not endangered. Risk, on the other hand, depends on how those hazards could affect human activities. In general, geologic hazards can't be controlled by humans. However, we can do much to minimize risks through wise land use, timely warnings, and community preparedness.

The idea of **acceptable risk** means being willing to live with a certain amount of risk in exchange for some economic or social benefit. Government agencies often decide how much risk is acceptable, balancing risks with economic, social, and political factors. For example, zoning laws may limit population density or restrict hospital construction in hazardous areas.

Most geological events occur in sparsely populated areas, and death tolls are low. Even small events can have deadly results, however, if they strike populated areas without warning. The worst situation is that of a large, unexpected event in a densely populated area.

History is the key

Depending on its size, type, and duration, a geologic event can affect areas up to hundreds of kilometers from its source. In the case of tsunamis, damaging effects can be global in extent. We can identify areas most likely to be affected by geologic hazards through a detailed study of an area's geologic history. Knowing the type and extent of previous activity, the geologic setting, population density, and local construction practices, we can identify hazardous zones.

Recurrence intervals

If tectonic plates move steadily, earthquakes and volcanic eruptions should occur fairly regularly, given that the plate motions cause these events.

A recurrence interval is the time expected between two similar geologic events generated by the same geological feature. For example, if a magnitude 6 earthquake occurs in the same area five times in 100 years, it has a recurrence interval of 20 years (100 years ÷ 5 earthquakes = 20 years per earthquake).

Although the historical record of earthquakes and volcanic eruptions is incomplete, it allows us to estimate how frequently major events occur in an area. By looking at the past, geologists can make educated guesses about what to expect in the future.

Unfortunately, people tend to view risk in terms of their lifetimes. This can be dangerous, because recurrence intervals for major earthquakes and volcanic eruptions can range anywhere from 20 years to as long as 20,000 years.

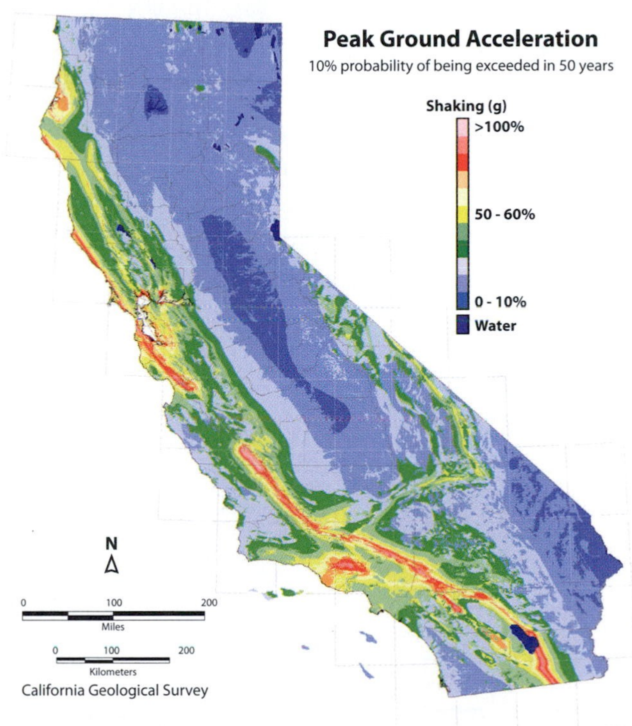

This seismic risk map of California shows areas with a 10-percent probability of experiencing ground shaking in the next 50 years. The color scale indicates varying intensities of horizontal acceleration (shaking) of the ground, relative to vertical acceleration of an object due to gravity (g). Geologic risk maps often include information about the intensity, extent, and expected frequency of a particular type of hazard event.

Seismic risk factors

Every day, hundreds of earthquakes occur somewhere on Earth. These earthquakes form identifiable patterns that outline tectonic plate boundaries. A significant number of these events occur on or near heavily populated coastlines. Some large earthquakes kill or injure thousands, while others go virtually unnoticed. So, what makes an earthquake deadly?

Three main factors influence how devastating an earthquake will be: (a) the intensity of ground shaking, (b) human factors, such as the population density of the region, and (c) economics or the wealth of the region.

Intensity of ground shaking

Ground shaking is controlled by the magnitude, depth, and distance of the earthquake; the direction of slip on the fault; and local conditions such as soil type and topography.

Earthquake magnitude scale

Magnitude	TNT for seismic energy yield	Example	Earthquake effects
-1.5	6 ounces	Breaking a rock on a lab table	Generally not felt but recorded by seismographs
1	30 pounds	Large blast at a construction site	
2	1 ton	Large quarry or mine blast	
3	29 tons		Often felt but damage rare
4	1000 tons	Small nuclear weapon	
5	32,000 tons		Minor/moderate damage
6	1 million tons	Double Spring Flat, NV quake, 1994	Can be destructive in areas up to 100 kilometers wide
6.5	5 million tons	Northridge, CA quake 1994	
7	32 million tons	Hyogo-Ken Nanbu, Japan quake, 1995	Major earthquake. Can cause serious damage over larger areas
7.5	160 million tons	Landers, CA quake, 1992	
8	1 billion tons	San Francisco, CA quake, 1906	Great earthquake. Can cause major damage in areas hundreds of kilometers wide
8.5	5 billion tons	Anchorage, AK quake, 1964	
9	32 billion tons	Chilean quake, 1960	
10	1 trillion tons	San Andreas-type fault circling Earth	Theoretical quake (equal to Earth's daily "dose" of solar energy)
12	160 trillion tons	Fault through center of Earth	

Source - http://www.seismo.unr.edu/ftp/pub/louie/class/100/magnitude.html

In 1935 Dr. Charles Richter created a scale for measuring earthquakes in southern California. By measuring the amplitudes of seismic waves recorded on his seismograph, Richter was able to estimate the earthquake's energy. Although Richter's scale applies only to earthquakes recorded in southern California, his work led to the development of a magnitude scale that can be used worldwide.

Magnitude—When rock under stress suddenly breaks and moves during an earthquake, it releases a tremendous amount of energy. Magnitude describes the size or strength of an earthquake at its point of origin. The magnitude increases when more rock is broken and slips along a fault. The amount of energy that is converted into seismic waves also increases with magnitude. These waves cause the ground shaking and damage that occur during earthquakes.

Depth and distance—Magnitude is not the whole story. As waves move away from the earthquake, their size and energy decrease. If an earthquake occurs at great distance or depth (>500 km away or below the surface) the waves generally cause little damage.

Near the earthquake, movement along a fault may rupture the ground surface. In addition to cracking building foundations and roads, surface ruptures can break gas, water, sewer, and communication lines. This creates additional health and safety hazards and hampers rescue efforts.

Soil type—Another critical factor in earthquake risk is the type of soil on which structures are built. Shaking in hard rock dissipates quickly, whereas loosely compacted sediments amplify seismic waves and intensify the ground shaking.

This process was dramatically illustrated during the 1985 Mexico City earthquake. The earthquake actually occurred 320 km west of the city, near the Pacific coast. There, the bedrock is very hard, so only minor damage and few deaths occurred. However, in Mexico City, which is built atop the soft sediments of an old lakebed, around 10,000 people were killed. The shaking triggered by the magnitude 8.1 quake was amplified by the loose soils, causing buildings to collapse or settle.

Karl V. Steinbrugge, Courtesy National Information Service for Earthquake Engineering, University of California, Berkeley

One of many buildings that collapsed in the magnitude 8.1 earthquake on September 19, 1985 in Mexico City.

Soil conditions also played a major role in the 1989 Loma Prieta earthquake that affected San Francisco. An area of the city known as the Marina District sits on the edge of San Francisco Bay. The combination of loose fill and the high water table beneath this area enhanced the seismic shaking. The Marina District was one of the hardest hit areas of the city in the 1989 quake.

Human factors

Population density—An earthquake, no matter how powerful, doesn't pose a direct risk to human lives unless it occurs near where people live. The map below shows the population density of California. Darker areas have a high density, whereas lighter areas have a lower density. Much of the population is clustered along the Pacific Coast, where San Diego, Los Angeles, and San Francisco are located. Unfortunately, the coast is also the area where seismic activity occurs most frequently. (Compare it with the seismic risk map on page 67).

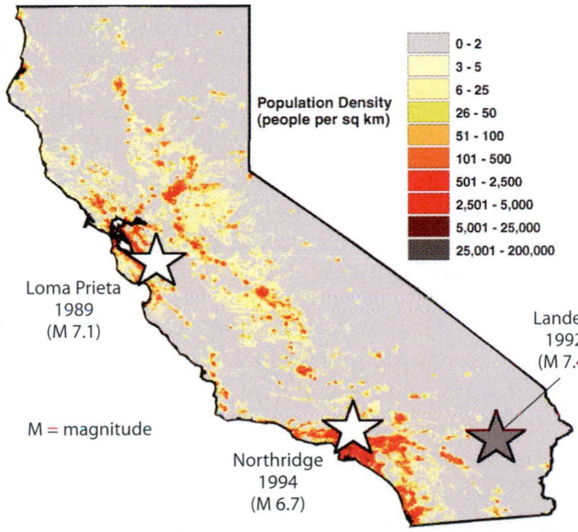

When strong earthquakes occur where the population density is high, the effects can be catastrophic. Three recent California quakes illustrate the effect of population density on the overall impact of an earthquake. The two white stars in the lower-left map represent the Loma Prieta and Northridge earthquakes. Together, these quakes killed almost 120 people and caused over 20 billion dollars in damage. The gray star shows the location of the Landers earthquake, the most powerful quake to hit California since 1952. Because of the low population density, the quake did only 100 million dollars in damage and killed just one person.

Human activities—Another critical factor that influences the number of people affected by earthquakes is where people are and what they are doing at the time an earthquake strikes. People who live in wood-frame homes, for example, may be more protected from earthquakes that occur in the middle of the night than those who live in tall, poorly-constructed buildings made of unreinforced concrete.

J.K. Nakata, USGS

In San Francisco's Marina District, nine people died in 1989 in building collapses and fires. This automobile was crushed under the third story of an apartment building. The first two floors of the structure are no longer visible.

The 1989 Loma Prieta earthquake struck just as a World Series baseball game between the two Bay Area teams was getting under way. Fortunately, many people who might otherwise have been driving home from work were watching the game. That lucky break probably saved hundreds of lives, because a large portion of a freeway normally packed with rush hour traffic collapsed during the earthquake. The upper deck of the double-decker highway fell onto the lower deck, killing 42 people.

The timing of an earthquake can also be disastrous. The January 26, 2001 Bhuj earthquake in India happened on Republic Day, a national holiday. When the quake struck, many citizens were in the streets, participating in or watching parades. Buildings collapsed onto the spectators, killing hundreds who might have otherwise been safe.

H. G. Wilshire, USGS

This section of Interstate 880 in Oakland, California collapsed in the 1989 Loma Prieta earthquake. Support columns in the two-tiered freeway failed, causing sections of the upper deck to pancake onto the lower deck.

Economic factors

Gross Domestic Product (GDP) represents the total value of goods and services a country produces in a year. Usually, a country's GDP is divided by its total population to obtain the GDP per person, or per capita. GDP per capita provides a good estimate of a country's wealth relative to other countries.

A country's wealth has a dramatic effect on its ability to plan for, endure, and recover from the effects of natural disasters. Affluent countries can afford to enforce stringent building codes and use superior construction materials. This reduces the chances of building collapse, which killed tens of thousands in recent quakes in Turkey (1999) and India (2001).

Wealthier countries also have more modern search-and-rescue and health-care systems to locate and treat the injured following earthquakes. Insurance networks and disaster-relief programs help citizens and their communities rebuild after earthquakes.

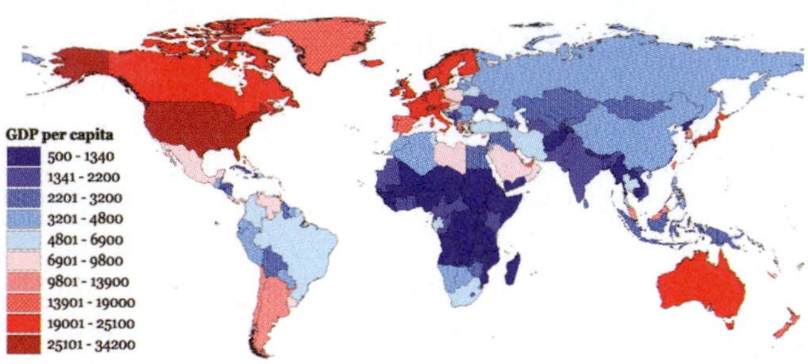

The world's per capita distribution of Gross Domestic Product. Wealthier countries are shown in red, poorer in blue.

Building construction is a major factor in determining an earthquake's destructive capability. In earthquake-prone regions, buildings must be able to withstand the horizontal shaking of an earthquake to prevent their collapse. Buildings should have strong connections between the walls, ceilings, and floors to keep them from breaking apart.

Some of the most earthquake-safe buildings are single-story, wood-frame houses with strong connections between the roof, walls, and foundation. Wood is flexible and can move with the sway of the building. It may be damaged, but it generally does not collapse unless the roof is extremely heavy.

Concrete and brick buildings must be reinforced with steel to keep them from cracking and breaking apart. Steel is flexible yet strong and will bend rather than break like concrete. On the other hand, failure in steel buildings most often occurs at weak welds between parts of the building.

Lindie Brewer, USGS

This sturdy wood-frame house was located just 18 meters from the Landers earthquake fault scarp. While heavily damaged, the house did not collapse and its occupants, though shaken, escaped unharmed.

Building damage often depends on the frequency of the ground motion — that is, how fast the ground vibrates. Damage can be particularly severe if the frequency of ground motion matches the natural vibration frequency of the structure, causing the building to resonate.

Resonance occurs when you push a child on a swing. If you push at the right time, the swing goes higher and faster. If you push at the wrong time, the swing slows down and ultimately stops. When seismic waves shake a building at its natural frequency, the building resonates and can quickly self-destruct.

Tall buildings, bridges, and other large structures respond more to low-frequency shaking, whereas small structures respond more to high-frequency shaking. During the 1985 Mexico City earthquake, most of the buildings that collapsed were between 7 and 15 stories tall. The low-frequency ground vibrations of the earthquake matched the natural vibration frequency of

E. V. Leyendecker, National Bureau of Standards

This steel frame building in Mexico City collapsed due to resonant ground shaking. The taller concrete building behind it suffered little damage.

buildings in this height range. Buildings shorter than 7 stories or taller than 15 stories sustained far less damage.

Other hazards

Fire—Broken gas and power lines often start fires after major earthquakes. These fires can easily spread out of control, particularly if broken water mains limit or cut off the water supply. Fires contributed significantly to damage and deaths following the 1906 and 1989 San Francisco earthquakes and a 1995 earthquake in Kobe, Japan.

Landslides—In mountainous areas, earthquakes can trigger landslides. The sudden and total devastation caused by landslides can produce extremely high death tolls. On May 21, 1970, a magnitude 7.9 earthquake on Mt. Huascarán, Peru triggered a rock and snow avalanche that buried the towns of Yungay and Ranrahirca. In all, the landslide killed over 20,000 people and caused more than $250 million dollars in damage.

Tsunamis—Earthquakes occurring in or near oceans can create seismic sea waves, or ***tsunamis***. A large tsunami can travel great distances across an ocean with a minimal loss of energy. Thus, tsunamis are capable of causing destruction far from their source. Tsunamis and their hazards are covered in more detail in the Tsunami Hazards activity.

Hazardous spills—The growing use and storage of hazardous materials increases the potential risk from earthquakes. Damage to storage tanks, pipelines, and transportation systems for these materials could have catastrophic results that last well beyond the immediate effects of an earthquake.

G. Plafker, USGS

The 1970 Mt. Huascarán landslide consisted of nearly 100 million cubic meters (80 million cubic yards) of water, mud, and rocks. Almost 68,000 people were killed by combined effects of the earthquake and landslide.

NOAA/NGDC

Panoramic view of San Francisco on April 18, 1906 shows the city in flames following the magnitude 8.4 earthquake. The death toll was more than 3,000 from all causes, and damage was estimated at $500 million in 1906 dollars (about $10 billion in 2004 U.S. dollars) .

Questions

1. What is the difference between a geologic hazard and a geologic risk?

2. California and Alaska both have very strong earthquakes. Why might the earthquake risk be much higher in California than it is in Alaska?

3. If a major earthquake occurs four times on a segment of a fault in 100 years, what is the recurrence interval of major earthquakes for that fault segment?

4. Why was so much damage done in Mexico City in 1985 by an earthquake that originated 320 km (200 miles) away?

5. Why was the death toll from the January 2001 earthquake in Bhuj, India unusually high?

6. What is "GDP per capita," and how might it affect the number of deaths in an earthquake?

7. Would you prefer to be in a modern, steel-frame skyscraper; a 3-story, brick office building; or a 1-story, wood-frame house during an earthquake? Explain why.

8. Why, in some earthquakes, does most of the damage occur to buildings of a certain height, whereas taller and shorter buildings receive little damage?

Investigation 3.4

Seismic risk and society

In this activity, you will explore the factors that influence seismic hazard and seismic risk around the world.

- Launch ArcMap, and locate and open the **etde_unit_3.mxd** file.

Refer to the tear-out Quick Reference Sheet located in the introduction to this module for GIS definitions and instructions on how to perform tasks.

- In the Table of Contents, right-click the **Earthquake Hazards** data frame and choose Activate.
- Expand the **Earthquake Hazards** data frame.
- Turn off the **Earthquakes** and **Topography** layers.
- Turn on the **Seismic Hazard** layer.
- Select the **Seismic Hazard** layer.

How likely is it?

When talking about hazards and risks, we usually talk about ***probability***—how likely it is that some event will happen in a given period of time. In this case, the seismic hazard is defined as a 10 percent probability of some maximum ground acceleration over 50 years. According to the seismic hazard legend, an area classified as a Very High seismic hazard is rated at over 40 percent g. Thus, such an area has a 10 percent probability of experiencing an earthquake with shaking of 40 percent g (3.9 m/s^2) or greater in the next 50 years.

The **Seismic Hazard** layer shows the maximum shaking, or ground acceleration, that an area could experience over the next 50 years. There is only a 10 percent probability that the maximum level of shaking will occur, so the map represents the worst case expected for the area. The hazard units are expressed as percentages of **g**, the acceleration of gravity (9.8 m/s^2). The greater the ground acceleration, the more likely there will be structural damage and, as a result, injuries and deaths.

Seismic hazards, plate tectonics, and fault systems

As you've seen in your investigation of earthquakes and plate boundaries, large earthquakes are often associated with plate margins. Next, you will determine if this relationship holds true for earthquake hazards.

- Click the Select By Attributes button.
- To display areas of moderate to very high seismic hazard, query the **Seismic Hazard** layer for **("Hazard" = 'Moderate') or ("Hazard" = 'High') or ("Hazard" = 'Very High')** as shown in steps 1–6 on the following page. The query will actually read:

("HAZDESC" = 'Moderate') OR ("HAZDESC" = 'High')
OR ("HAZDESC" = 'Very High')

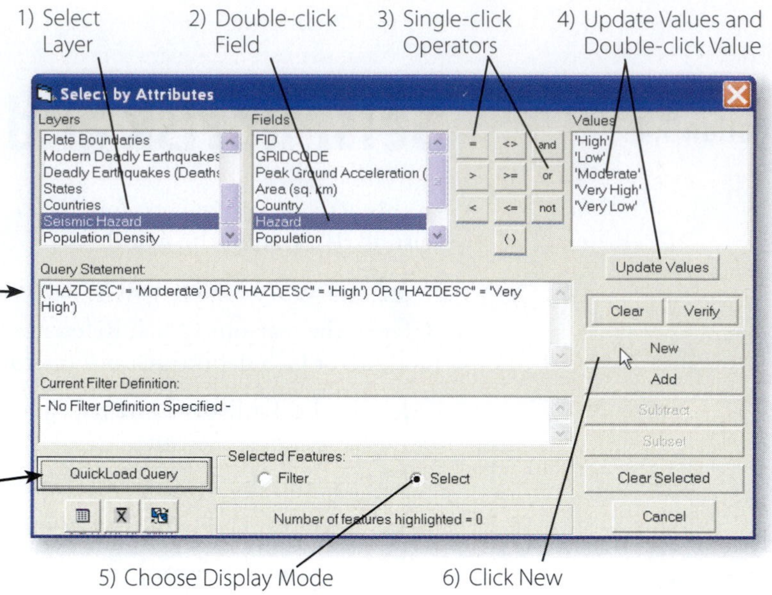

Read query statement here as you enter it.

QuickLoad Query

- Click the QuickLoad Query button and select the **Seismic Hazard** query.
- Click **OK**.
- Click **New**.

- If you have difficulty entering the query statement correctly, refer to the **QuickLoad Query** described at left.
- Close the Select By Attributes window.

Now the areas of moderate to very high seismic hazard are highlighted on your map.

- Turn on the **Plate Boundaries** layer.
- Use the Zoom In and Zoom Out tools to examine seismic hazard areas near the plate boundaries. (You may need to zoom in close to examine these areas.)

1. Which plate boundary type is most commonly associated with areas of moderate to very high seismic hazard?

Seismic hazards far from plate boundaries

Not all seismic hazards are found near plate boundaries.

- Click the Clear Selected Features button.
- Turn on the **Unusual Seismic Hazards** layer.
- Select the **Unusual Seismic Hazards** layer.

This layer shows areas where seismic hazards exist far from recognized plate boundaries.

- Click the Identify tool.
- In the Identify Results window, select the **Unusual Seismic Hazards** layer from the list of layers.
- Click in each of the five unusual seismic hazard regions. Read the information in the Identify Results window, then close the window. Use this information to complete the table on the following page.

2. Complete the following table by filling in the name of the region that matches each seismic hazard source.

m.y. = million years ago

Region	Probable origin of seismic hazard
	Collision of two continents beginning 40 million years ago. The broad zone of deformation is due to the fact that both continents are made of low-density continental rock that cannot be subducted or destroyed.
	Tsunamis (seismic sea waves) originating at local plate boundaries, and internal deformation of the plate related to ancient tectonics that formed the continent.
	Rifting — the continent is breaking apart, or rifting, along a newly forming plate boundary.
	Glacial rebound — continents are rising in response to ice sheets that melted 11,000 years ago. This causes small to moderate earthquakes.
	Seismic activity is related to weaknesses in the plates at ancient plate boundaries.

Glacial rebound

During the last ice age, around 20,000 years ago, thick ice sheets covered much of the northern hemisphere. The weight of the ice made the continents sink deeper into the mantle. As the ice sheets melted, the continents rose, or rebounded.

Though most of the ice melted long ago, glacial rebound continues today in places such as Greenland, Iceland, and eastern Canada, producing small amounts of seismic activity.

Turn off the **Plate Boundaries** and **Unusual Seismic Hazards** layers.

Next, you will explore the difference between seismic hazard and risk.

From hazard to risk

Seismic hazard and seismic risk are related, but have very different meanings. Seismic hazard is the probability that ground shaking of a particular intensity will occur in a region, as determined from historic records. Seismic risk reflects society's ability to successfully endure an area's seismic hazard. In this section, you will investigate the relationship between seismic hazard and seismic risk. First, you will narrow your selection to only those areas where seismic risk is "Very High."

Select the **Seismic Hazard** layer.

Click the Select By Attributes button.

To display areas where seismic risk is very high, query the **Seismic Hazard** layer using the query statement **("Hazard" = 'Very High')**. The query will actually read:

("HAZDESC" = 'Very High')

Click the **New** button.

Now the areas of very high seismic hazard are highlighted on your map.

If you have difficulty entering the query statement correctly, refer to the **QuickLoad Query** described at left.

Close the Select By Attributes window.

QuickLoad Query

- Click the QuickLoad Query button and select the **Seismic Risk** query.
- Click **OK**.
- Click **New**.

Next, you will examine the characteristics of several countries located in zones of high seismic risk.

Click the Summarize button **Σ**.

In the Summarize window, select **Seismic Hazard** as the feature layer.

Select **Country Name** as the **field to summarize** in the drop-down menu.

- Double-click **Area (km2)** to display the statistics options and check **Sum** (but do not click **OK**).
- Double-click **Population** to display the statistics options and check **Sum** (but do not click **OK**).
- Double-click **GDP** to display the statistics options and check **Average**.
- Check the **Summarize on Selected Records only** option and click **OK**.

A new table called a *summary table* will open.

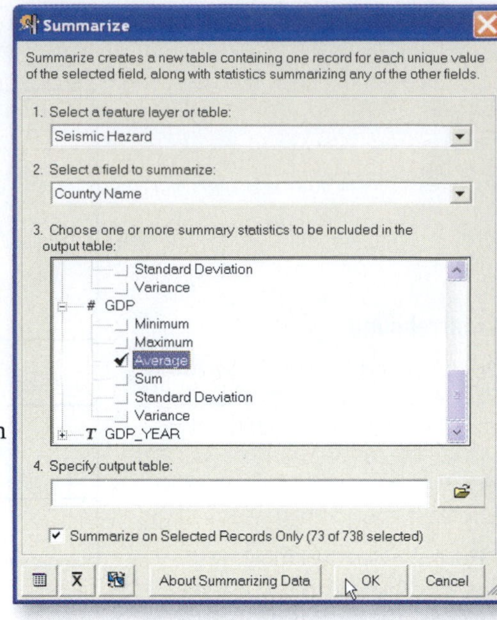

What is GDP?

Gross Domestic Product (GDP) is the total amount of money that changes hands in a country in a given year. GDP per capita is simply this total amount of money divided by the number of people living in the country.

3. Record the summary results in the table below. You will need to calculate the population density (see table for equation).

Country		Data for very high seismic risk zones		GDP per capita 2004 U.S. Dollars *Average GDP*
	Area *Sum KM2*	**Population** *Sum ZONEPOP*	**Population density** *Population ÷ Area*	
China				
Japan				
United States				
Turkey				
Iran				
Taiwan				
Pakistan				

Ranking countries within the highest risk areas

Each factor in the above table contributes to the overall seismic risk of a region. A large population or high population density increases the risk of death or injury from an earthquake. A country's wealth, as indicated by its Gross Domestic Product (GDP) per capita, is also important in determining seismic risk. Affluent countries are better able to plan for and respond effectively to earthquakes.

4. Using the data in the table above, rank the countries for each risk factor in the risk matrix in the following table (ranking them from 1 to 7 for each category). Calculate the Total risk ranking for each country by adding the risk values across each row.

Country	Population ranking *highest = 7* *lowest = 1*	Population density ranking *highest = 7* *lowest = 1*	GDP per capita ranking *lowest = 7* *highest = 1*	Total risk ranking
China				
Japan				
United States				
Turkey				
Iran				
Taiwan				
Pakistan				

Next you will compare this risk ranking to the effects of recent earthquakes in each of these countries.

The following table illustrates the deaths and damage caused by seven recent major quakes.

Country	Location	Year	Magni-tude	Deaths	Buildings destroyed	Damage *billion $U.S.*
China	Tangshan	1976	7.6	>240,000	180,000	5.6
Japan	Kobe	1995	6.8	5502	100,000	150
United States	Northridge	1994	6.7	57	7000	44
Turkey	Izmit	1999	7.4	19,000	35,000	6.5
Iran	Bam	2003	6.6	26,000	85%	0.033
Taiwan	Chichi	1999	7.6	2400	10,000	14
Pakistan	N. Pakistan/Kashmir	2005	7.6	>87,000	>32,000	5.2

5. How well does the risk factor you calculated in the risk matrix table compare to the actual earthquake damage and casualty figures for each of these countries? Discuss both the number of casualties and the total dollar damage.

Close the Summary Table when you are finished.

Earthquake deaths and wealth

A country's GDP per capita provides a good measure of its wealth. To complete the risk matrix table, we assumed that affluent countries are better prepared for earthquakes and better equipped to deal with their effects.

Now you will use ArcMap to test this assumption by looking at deadly earthquakes since 1950. You will examine how death and damage vary depending on whether the earthquake occurred in a wealthy country (GDP per capita >= $10,000) or a poor country (GDP per capita < $10,000).

- Click the Clear Selected Features button.
- Turn on the **Modern Deadly Earthquakes** layer.

This layer is a subset of the **Deadly Earthquakes** layer and shows the deadly earthquakes that have occurred since 1950.

- Click the Select By Attributes button.
- To display seismic hazard zones in countries where GDP per capita is at least $10,000 per year, query the **Seismic Hazard** layer using the following query statement:

  ("GDP" >= 10000)

- Click the **New** button.
- If you have difficulty entering the query statement correctly, refer to the **QuickLoad Query** described at left.
- Close the Select By Attributes window.

QuickLoad Query

- Click the QuickLoad Query button and select the **Wealthy Countries** query.
- Click **OK**.
- Click **New**.

The seismic zones located in wealthy countries are now highlighted on your map.

Now you will select the deadly earthquakes that are within 150 km of any of the seismic hazard zones in wealthy countries.

- Click the Select By Location button.
- In the Select By Location window, construct the query statement:

  I want to **select features from** the **Modern Deadly Earthquakes** layer that **are within a distance of** the features in the **Seismic Hazard** layer.

- Check the **Use selected features** box.
- Apply a buffer of **150 km** to the selected features of the **Seismic Hazard** layer.
- Click the **Apply** button.
- Close the Select By Location window.

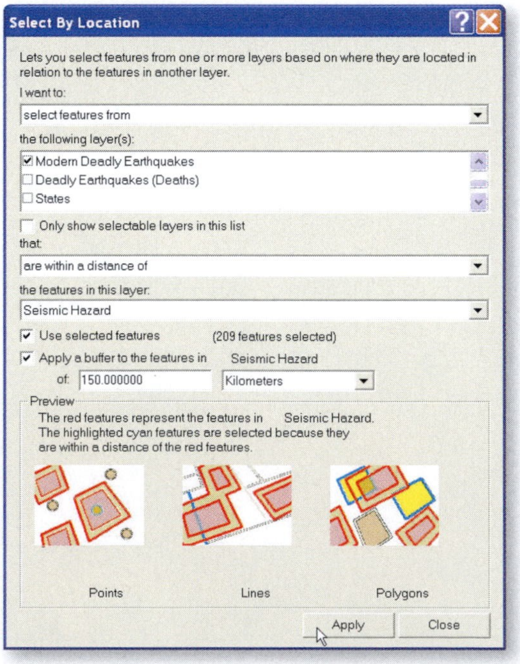

To determine the average number of deaths per quake in wealthy countries:

- Click the Statistics button.
- In the Statistics window, calculate statistics for **only selected features** of the **Modern Deadly Earthquakes** layer, using the **Deaths** field.
- Click **OK**. Be patient while the statistics are calculated.

The average number of deaths per quake is given as the **Mean**.

6. What is the average number of deaths per quake in countries with a high GDP per capita (over $10,000)?

- Close the Statistics window.

To find the average number of deaths per quake in poor countries:

- Select the **Seismic Hazard** layer.
- Click the Switch Selected Features button. This will select all the seismic zones located in countries where GDP per capita is below $10,000 per year.
- Click the Select By Location button.
- In the Select By Location window, construct the query statement:

  I want to **select features from** the **Modern Deadly Earthquakes** layer that **are within a distance of** the features in the **Seismic Hazard** layer.
- Check the **Use selected features** box.
- Apply a buffer of **150 km** to the selected features of the **Seismic Hazard** layer.
- Click the **Apply** button.
- Close the Select By Location window.

To determine the average number of deaths per quake in countries with a GDP per capita less than $10,000:

- Click the Statistics button.
- In the Statistics window, calculate statistics for **only selected features** of the **Modern Deadly Earthquakes** layer, using the **Deaths** field.
- Click **OK**. Be patient while the statistics are calculated.

7. What is the average number of deaths per quake in countries with a low GDP per capita (under $10,000)?

8. Describe how a country's GDP might affect the number of injured or killed.

9. Summarize the factors that increase seismic risk in a country.

Close the Statistics window.

Quit ArcMap and do not save changes.

Unit 4
Volcano Hazards

In this unit, you will

- *Examine records of historical volcanoes.*
- *Use the Volcanic Explosivity Index (VEI) to categorize volcanoes.*
- *Investigate the effects of major eruptions on climate.*
- *Explore the effects of the most explosive volcanoes in history.*

Lyn Topinka, USGS

Mt. St. Helens viewed from the north, showing Spirit Lake and the new summit crater. The lateral blast from the May 18, 1980 eruption damaged or destroyed over four billion board feet of usable timber, enough to build 150,000 homes.

Warm-up 4.1

Location map

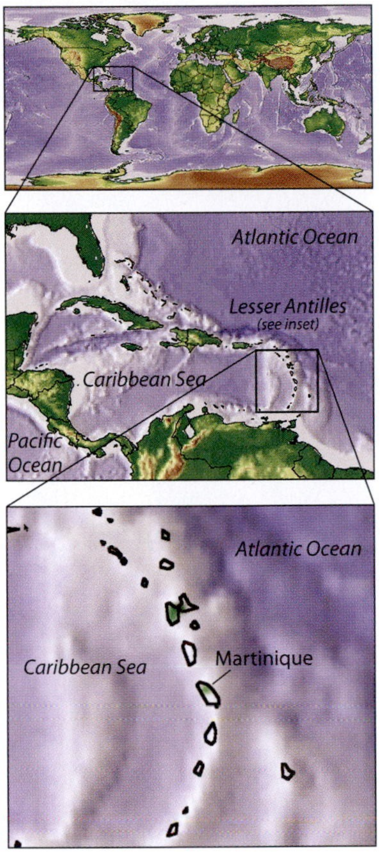

The tragedy of Mont Pelée

After reading the account of the volcanic eruption that destroyed the town of St. Pierre on the island of Martinique in 1902 (pages 85–89), list and describe all of the volcano-related hazards discussed. Feel free to add other hazards from your previous knowledge or experience with volcanoes.

Volcano hazards

1.

2.

3.

4.

5.

6.

7.

8.

Volcano hazard examples (optional)

If you have access to a computer, you can see examples of these and other volcanic hazards. Be sure to add any new hazards you find to your list.

- Launch ArcMap, and locate and open the **etde_unit_4.mxd** file.

Refer to the tear-out Quick Reference Sheet located in the introduction to this module for GIS definitions and instructions on how to perform tasks.

- In the Table of Contents, right-click the **Geological Hazards** data frame and choose Activate.
- Expand the **Geological Hazards** data frame.

To see examples of other volcano hazards:

- Turn on the **Hazard Links** layer.

- Select the **Hazard Links** layer.
- Using the Hyperlink tool, click on each of the brown volcano hazard symbols (brown volcanoes) on the map.
- Read the caption for each picture, then close its window. There may be more than one picture for each link.

Questions

1. In St. Pierre, which of the hazards you listed caused the greatest amount of damage and cost the most lives?

2. Do you think anything could have been done to reduce the loss of life and property in St. Pierre? Explain.

3. What warning signs did the people of St. Pierre have of the coming eruption? Why do you think so many people ignored the signs?

4. Why do you think people build farms and cities so close to volcanoes?

5. Would you rebuild the city of St. Pierre? Explain why or why not.

Eyewitness accounts

The Destruction of St. Pierre, 1902

Volcanoes in the Caribbean

The islands of the Lesser Antilles are the surface expression of a volcanic arc above a subduction zone. For the past 40 million years, the Atlantic plate has been slowly plunging beneath the Caribbean plate, building this chain of tropical islands.

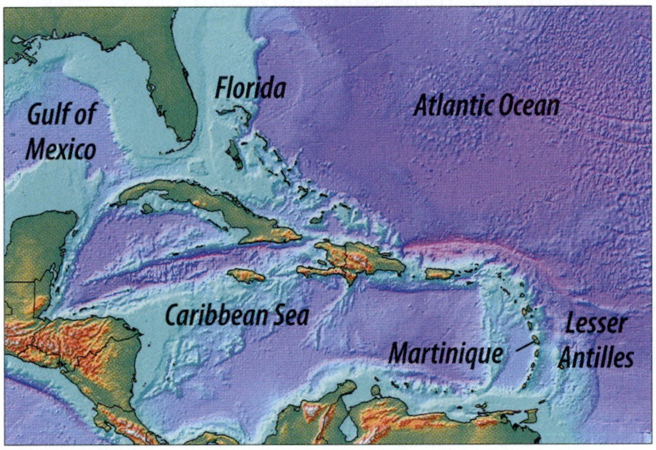

The rock and sediment of the ocean floor contain significant amounts of water. As the Atlantic plate carries these rocks downward into the mantle, the increasing heat and pressure releases the trapped water. At a depth of about 100 km (60 mi), this combination of heat, pressure, and water vapor cause the subducting plate to melt. The molten rock — **magma** — is much lighter than the surrounding mantle, so it rises toward the surface. Near the surface, the magma accumulates in reservoirs known as magma chambers. Occasionally, the pressure in the chamber increases, forcing magma to the surface in a volcanic eruption.

The shape and explosive nature of these volcanoes is a direct result of the magma's chemical content. At subduction zones, minerals with relatively low melting points such as silica (quartz) melt, whereas minerals with higher melting points remain solid. Magmas high in silica are very viscous; that is, they do not flow easily. Rather, the water vapor and other gases in the magma tend to build pressure until it is released explosively. This type of eruption forms steep-sided cones called **stratovolcanoes** that can reach thousands of meters in elevation. Mt. Rainier (outside Seattle), Mt. Fuji (in Japan), Popocatépetl (towering over Mexico City), and Mt. Etna (on the island of Sicily) are all stratovolcanoes, and all are located above subduction zones.

The tropical islands of the West Indies are a series of stratovolcanoes that occasionally erupt with spectacular violence. The first recorded eruption in the West Indies occurred about 1660 on the island of Montserrat. Since that time, dozens of eruptions have been observed and activity continues with regularity today. The deadliest eruption of the 20th century occurred in the West Indies on the island of Martinique.

Mt. Pelée and the island of Martinique

The island of Martinique sits at the midpoint of the Lesser Antilles island arc. For several hundred years the island had been a stopping point for pirates and buccaneers, but by the start of the 20th century it was home to over 160,000 residents. The largest city, the port of St. Pierre, was a thriving community of over 25,000. Spread out along a gently curving beach at the foot of cone-shaped Mont Pelée, it billed itself as the "Paris of the West Indies." Famous for its rum, St. Pierre had a distillery filled with barrels of the potent spirit.

Mont Pelée, rising several miles to the north of St. Pierre, was draped in a thick cover of jungle. At its summit, a bowl-shaped crater held a lake, *Lac des Palmistes*. Below the summit was a second crater, breached by a V-shaped canyon aimed like a rifle sight at St. Pierre, 6 km (3.8 mi) away. Though it was well known that Mont Pelée was a volcano, having produced minor eruptions in 1792 and 1851, the 1300-m (4265-ft) peak was assumed by most to be dormant and therefore harmless. In fact, it was considered by many to be the island's benevolent "protector."

Mt. Pelée awakens

In late January 1902, Mont Pelée began to show signs that it was awakening from its long slumber. Smoke and steam began to puff from the summit. For the people of St. Pierre, it was more a source of wonder than alarm. This activity, called **fumarole** activity, gradually increased through the spring. Citizens occasionally noted the smell of sulfur, and on many days the mountain's summit was covered with an ashen fog.

A. Lacroix

A later cloud of hot gas and debris (*nuée ardente*) roars down the side of Mont Pelée in June 1902, passing over the already destroyed city of St. Pierre.

On April 23rd the townsfolk heard a loud explosion, and the summit of Mont Pelée billowed gray ash skyward in a minor eruption. Over the next few days, there were numerous explosions and tremors, and several times St. Pierre was dusted with fine volcanic ash. Conditions for the citizens of the town deteriorated; several times clouds of sulfurous gases, smelling of rotten eggs, permeated the city. The increasing activity of Mont Pelée caused the wildlife of the mountain to seek safer surroundings, while deadly snakes and swarms of insects invaded St. Pierre and surrounding villages in search of food. Reports from diaries and letters described how livestock screamed as red ants and foot-long centipedes bit them. Many livestock and an estimated 50 people, mostly children, died from snakebites.

On May 5th the water in the crater lake was heated to a boil and the crater rim failed. The hot water rushed down the canyon towards St. Pierre, mixing with the recently fallen ash to create a volcanic mudflow called a **lahar**. The lahar destroyed everything in its path, including the premiere distillery where 23 workers were swept away and killed. When the lahar reached the ocean it created a local tsunami that flooded low-lying areas around the waterfront.

Obviously, life in the shadow of Mont Pelée was becoming unbearable. People were trying to leave the island, but passage aboard ships was difficult to secure. The cathedral was crowded with people waiting to make confessions. Seeking to comfort the populace, the governor sent a team to the summit to assess the danger. Only one scientist, the local high school science teacher, was in the group. The team delivered a positive report to the governor stating, "There is nothing in the activity of Mt. Pelée that warrants a departure from St. Pierre," and concluded that "the safety of St. Pierre is absolutely assured."

Unfortunately, the report was not accurate. On May 7th Pelée continued to be rocked with explosions, and people noted that two fiery "eyes" appeared near the summit. Above the summit hung a gray ash cloud filled with lightning. Marino Leboffe, captain of the Italian merchant ship *Orsolina*, knew what was coming. His home port was Naples, which lies in the shadow of Mt. Vesuvius, a volcano that erupted with deadly force in A.D. 79, completely burying the cities of Pompeii and Herculaneum. Leboffe ordered the *Orsolina* to leave port only half loaded with cargo and noted, "I know nothing of Mont Pelée, but if Vesuvius were looking the way your volcano looks this day, I'd get out of Naples; and I'm going to get out of here." Threatened with arrest for leaving port without clearance papers, Leboffe replied, "I'll take my chance of arrest, but I won't take any chances on that volcano."

The eruption on May 8th

Shortly before 8:00 a.m. on May 8th a series of violent explosions rocked Mont Pelée. Thus began a huge volcanic eruption, first sending a plume of gases and ash skyward, followed by a lateral eruption that sent a deadly superheated cloud of gases through the V-notch directly towards St. Pierre. The eruption that went upward created a huge mushroom cloud that blocked out the early morning sun and was pierced by flashes of lightning. People 30 km from the volcano were immersed in darkness, and could not see even an arm's length away. This cloud reached an elevation of nearly 9 km (30,000 ft) in a matter of a few minutes.

William E. Scull

An artist's depiction of the May 8, 1902 eruption.

Although spectacular, the vertical plume of ash was not the deadliest messenger from Mont Pelée. The lateral eruption of superheated gases roared toward St. Pierre at over 160 km per hour (100 mph). The shock wave flattened brick buildings and ripped branches from trees. "Rubble walls three feet in thickness had been torn to pieces as if made of dominoes or kindergarten blocks." Not a single roof remained attached. It was reported that a 3-ton cast-iron statue was thrown 15 m (50 ft) from its pedestal. The **nuée ardente**, or "fiery cloud" of superheated gases and debris, glowed red, searing everything in its path and igniting fires throughout the city. At the distillery, rocks propelled by the cloud pierced centimeter-thick iron storage tanks with holes as large as 30 cm (1 ft) in diameter. Thousands of barrels of rum exploded, sending flaming rivers down the streets of the city and out to sea.

The eruptions were immediately followed by torrential rains. The runoff mixed with volcanic ash to form lahars that swept down river valleys, filling them with debris and burying or sweeping houses off their foundations. The lahars flowed into the sea so suddenly that they generated large waves — tsunamis — that were observed around the Caribbean.

When the eruption finally subsided, more than 29,000 people had been killed. Only two residents of St. Pierre were reported to have survived, making this eruption the deadliest of the 20th century.

Stories of survivors

The force of the eruption was not limited to the city. In the harbor, the shock wave capsized steamships, and raining pyroclastic debris set ships aflame. Chief Officer Ellery Scott of the Canadian steamship *Roraima* later told about his experience. Below is a translation of his account.

Chief Officer Ellery Scott

According to Scott, "The ship arrived at St. Pierre at 6 a.m. on the 8th. At about 8 o'clock, loud rumbling noises were heard from the mountain overlooking the town, the eruption taking place immediately, raining fire and ashes; lava running down the mountainside with a terrific roar, sweeping trees and everything in its course. I went at once to the forecastle-head to heave anchor. Soon after reaching there, there came a terrible downpour of fire, like hot lead, falling over the ship and followed immediately by a terrific wave which struck the ship on the port side, keeling her to starboard, flooding ship, fore and aft, sweeping away both masts, funnel-backs and everything at once.

"I covered myself with a ventilator standing nearby, from which I was pulled out by some of the stevedores, and dragged to the steerage apartment forward, remaining there for some time, during which several dead bodies fell over and covered me. Shortly after, a downfall of red hot stones and mud, accompanied by total darkness, covered the ship. As soon as the downfall subsided, I tried to assist those lying about the deck injured, some fearfully burnt. Captain Muggah came to me, scorched beyond recognition. He had ordered the only life boat left to be lowered; but it was too badly damaged. From that time, I saw nothing of the captain; but was told by a man that the captain was seen by him to jump overboard. The man followed him in the water, and succeeded in getting the captain on a raft floating nearby, where he died shortly after.

A. Lacroix

The ruined city of St. Pierre, as it appeared on February 19, 1903.

"I gave all help possible to passengers and others lying about the deck in dying condition, most of whom complained of burning in the stomach. I picked up one little girl lying in the passageway dying, covered her over with a cloth, and took her to a bench nearby, where I believe she died. About 3 p.m. a French man-of-war's boat, the *Suchet*, came alongside and passed over the side about twenty persons, mostly injured, and myself and other survivors were taken to Fort de France. I afterwards saw the *Roddam* steaming out to sea, with her stern part on fire. The *Roraima* caught fire and was burning when I left her in the afternoon, the town and all shipping destroyed."

Overall, 48 of the 68 crew members and passengers died in the horrible ordeal, while on other ships the casualty rate was even higher.

Ciparis

Although only two people survived the eruption in St. Pierre, several people lived in the surrounding communities and chronicled some remarkable tales. The most amazing story was that of Louis Auguste Sylbaris, also known as Ciparis. Ciparis was a robust 25-year-old who had a passion for drinking and brawling. In early April, he was arrested and jailed for wounding one of his friends with a sword during an argument. Jail time meant that Ciparis was required to labor in the service of the city, and he soon tired of the regimented life. Near the end of April, he escaped and partied all night with friends.

In the morning he turned himself in to the authorities, who sentenced him to one week of solitary confinement in the dungeon. The dungeon was next to a steep hillside, totally isolated from the outside world. On the morning of May 8, while Ciparis was waiting for his breakfast to be delivered, his cell became very dark. Gusts of hot air and steaming ash came through the cracks in his cell door. The heat became unbearable and he held his breath as long as he could. Finally, the heat began to subside and he slumped to the floor. He was horribly burned, but had some water in the cell to drink.

Ciparis managed to survive for four days before being rescued. An American journalist interviewed him shortly after his rescue and stated, "He had been more frightfully burned, I think, than any man I had ever seen." Ciparis eventually recovered, and was pardoned for his crimes on the basis of the miracle of his survival. Later, he joined the Barnum and Bailey Circus, and he was billed as the "Lone Survivor of St. Pierre."

Aftermath of the eruption

News of the deadly eruption was telegraphed to the world, and soon aid and supplies were rushed to Martinique. Along with this aid came scientists to study the volcano including Alfred Lacroix, often called the "father of modern volcanology." It was Lacroix who coined the phrase *nuée ardente*, which he described as a "lateral blast propelled down-slope by gravity."

On May 20th, 1902 Mont Pelée exploded again with a giant eruption, probably larger than that of May 8th. No one died in this event, mostly because there was no one left. Throughout the summer and fall there were many small eruptions, and on August 3rd another *nuée ardente* destroyed the village of Morne Rouge southeast of St. Pierre, killing 1000 to 2000 people.

In October 1902 a lava dome began to rise out of the crater floor. This dome formed an imposing obelisk that has been described as the most impressive lava dome ever produced; it was 100–150 m (325 – 500 ft) thick at its base and soared to 358 m (1175 ft) above the crater rim. At times it rose at the remarkable rate of 15 m per day! This obelisk, nicknamed "the tower of Pelée," glowed an incandescent red at night until it finally became unstable and collapsed into a pile of rubble in March of 1903.

A. Lacroix

View of the obelisk, or spire, rising 358 meters above the crater rim.

Source — Morris, C. (1902). *The volcano's deadly work: From the fall of Pompeii to the destruction of St. Pierre.* (Washington, DC: W. E. Scull)

For a detailed account of the events surrounding the 1902 eruption, see *The Day the World Ended* by Gordon Thomas and Max Morgan Witts. (New York: Stein and Day, 1969.)

Investigation 4.2

Deadly volcanoes

Volcanoes have the potential for affecting much larger areas and numbers of people than earthquakes. Mudflows generated by an eruption can travel hundreds of kilometers, and ash can be carried by atmospheric currents all the way around the world.

In this activity, you will investigate different types of volcanic hazards and their effects on people and the landscape.

Launch ArcMap, and locate and open the **etde_unit_4.mxd** file.

Refer to the tear-out Quick Reference Sheet located in the introduction to this module for GIS definitions and instructions on how to perform tasks.

In the Table of Contents, right-click the **Volcano Hazards** data frame and choose Activate.

Expand the **Volcano Hazards** data frame.

Historical volcanoes

The **Volcanoes** layer shows the locations of all volcanoes that are known to have been active in the past 10,000 years. The **Volcanic Eruptions** layer contains information about known volcanic eruptions since 10,000 B.C. These data are primarily gathered through written historical accounts as well as field investigations by geoscientists. The data are reliable for those regions where written history was preserved and sparse where oral history was the tradition.

Use the Zoom In tool and the Pan tool to closely examine the relationship between volcanoes and plate boundaries.

1. With which type of plate boundary are the volcanoes most strongly associated?

Click the Full extent button to view the entire map.

Turn off the **Volcanoes** and **Plate Boundaries** layers.

Select the **Volcanic Eruptions** layer.

Click the Open Attribute Table button.

Read the total number of volcanic eruptions at the bottom of the table (your answer will be different than the example shown below).

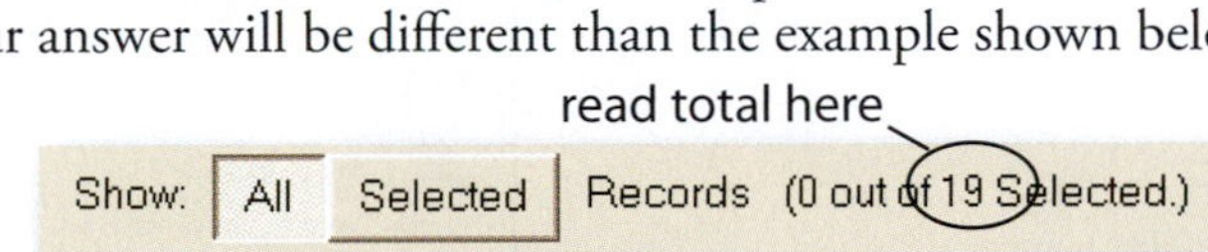

2. How many volcanic eruptions have been recorded since 10,000 B.C.?

Close the attribute table.

Volcanic Explosivity Index (VEI)

No single feature describes the size of a volcanic eruption. To compare the energy of eruptions, volcanologists developed a magnitude scale called the Volcanic Explosivity Index (VEI). The calculation factors in the height of the eruption plume, the distance ejected materials traveled, the duration of the blast, and the volume of material erupted. A VEI difference of 1 represents a difference in energy of approximately 10 times. For historical eruptions, VEIs range from 0 – 7, with 7 being the most explosive.

Now you will examine the largest historical eruptions.

Tephra — the solid material that is ejected into the air during a volcanic eruption.

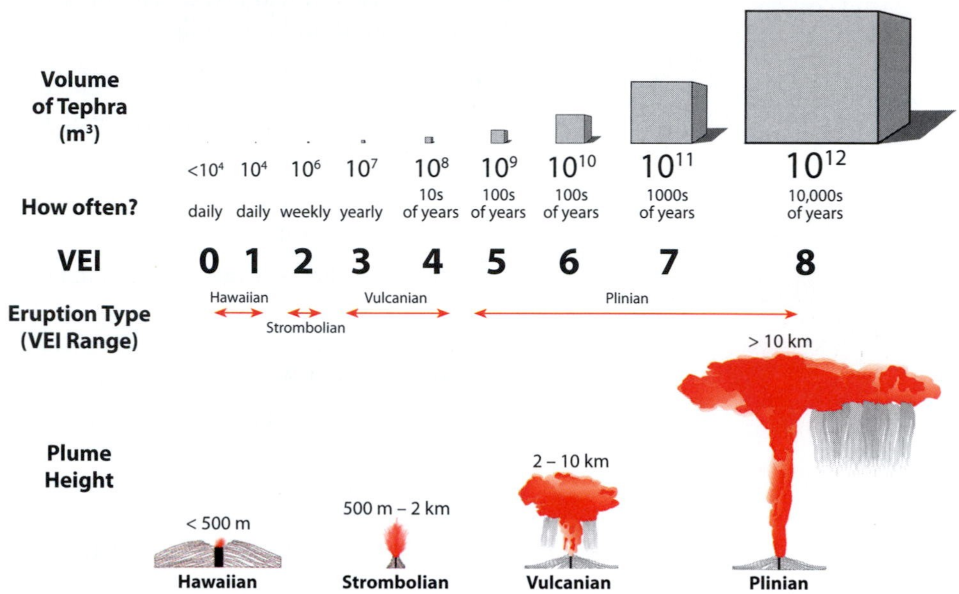

Which one's the biggest?

The only eruption in recent history with a VEI of 7 was the April 18, 1815 eruption of Mt. Tambora in Indonesia. To see a table of other historic eruptions and their intensities, click the Media Viewer button and choose **VEI Index Table**.

- Click the Select By Attributes button.
- To display volcanic eruptions with a VEI 5 or greater, query the **Volcanic Eruptions** layer for **("VEI" > 4)** as shown in steps 1 – 6:

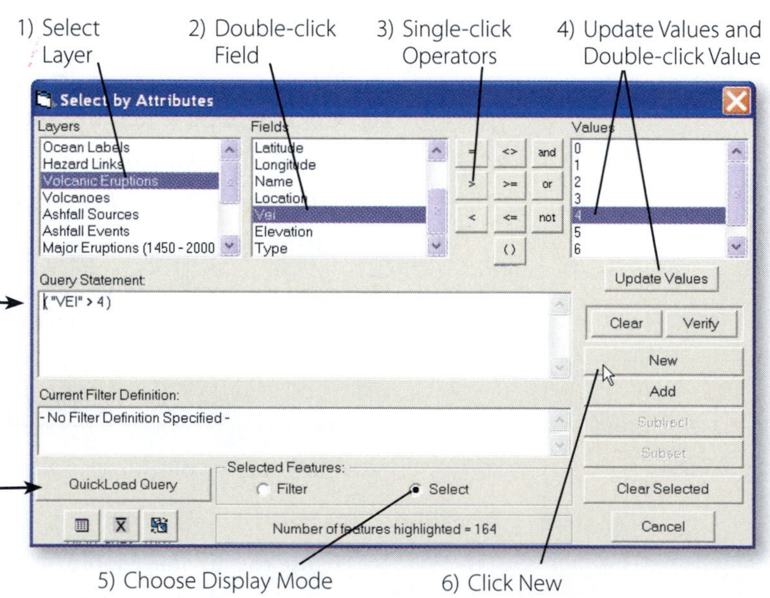

QuickLoad Query

- Click the QuickLoad Query button and select **VEI Query**.
- Click **OK**.
- Click **New**.

- If you have difficulty entering the query statement correctly, refer to the **QuickLoad Query** described at left on the previous page.
- Close the Select By Attributes window.

The volcanic eruptions with a VEI 5 or greater should now be highlighted on your map.

- Click the Open Attribute Table button.
- Read the number of major eruptions at the bottom of the table (your answer will be different than the example shown below).

3. How many major eruptions (VEI 5 or greater) have there been since 10,000 B.C.?

4. What percentage of the total eruptions were major ones (VEI 5 or greater)? Report your answer to the nearest tenth of a percent (see example at left).

To calculate a percentage

Divide the number selected by the total number and multiply by 100. For example, if 145 of 1300 eruptions were VEI 5 or greater, the percentage of eruptions of VEI 5 or greater would be

$$\% = (145 \div 1300) \times 100$$
$$= 0.112 \times 100$$
$$= 11.2$$

5. If an average of 250 volcanic eruptions occur each decade, how many eruptions would have a VEI greater than 4 in that same period?

- Click the Clear Selected Features button.
- Close the attribute table.

Historical records: a window to the future?

Scientists routinely use records of the past to make educated guesses about events in the future.

6. What factors might affect the reliability of historical data and our estimation of future hazards? Explain.

Recurrence interval—how often an event happens, or the number of years between similar events. Calculate this by dividing the total number of years by the number of events that occurred in that amount of time.

Historical data are often used to calculate recurrence intervals of major geologic hazards in a region. These are then used to estimate future levels of hazards. Recurrence can be calculated by dividing the number of years over which an event is repeated by the number of times the event occurred.

Next you will examine the recurrence interval of Mt. St. Helens eruptions with VEI ratings of 3 or greater.

- Click the Select By Attributes button.
- To determine the recurrence interval of VEI 3 eruptions or greater of Mt. St. Helens, query the **Volcanic Eruptions** layer using the following query statement:

  ("VEI" >= 3) AND ("NAME" = 'St. Helens')

- Click the **New** button.
- If you have difficulty entering the query statement correctly, refer to the **QuickLoad Query** described at left.

QuickLoad Query

- Click the QuickLoad Query button and select **Mt. St. Helens Query**.
- Click **OK**.
- Click **New**.

- Close the Select By Attributes window.
- Click the Open Attribute Table button.
- The number of eruptions found by your query is shown at the bottom of

the table (your answer will be different than the example shown below).

- Click the **Selected** button at the bottom of the table to view only the highlighted records. Use the table to answer the following questions.

7. How many times has Mt. St. Helens erupted with a VEI of 3 or greater? Over what time interval did those eruptions occur? (Hint: The database lists eruptions through the year 2004.)

8. What is the recurrence interval for eruptions of VEI greater than or equal to 3 by Mt. St. Helens? How would you use this information in managing the future development and use of the area surrounding the volcano?

- Close the attribute table.
- Click the Clear Selected Features button.

Range of volcanic hazards

On May 18, 1980 Mt. St. Helens erupted with a blast of VEI = 5 that deposited a blanket of ash over areas as far away as 400 km (250 mi). Lahars flowed down river valleys as far as 30 km (19 mi). Spokane, Washington was significantly affected by the event. What would be the impact of a similar eruption at Mt. Rainier near Seattle, Washington?

To figure this out:

- Turn off the **Volcanic Eruptions** layer and turn on the **Volcanoes** layer.
- Turn on the **Cities** layer.
- Select the **Cities** layer.
- Use the Zoom In tool to zoom in on the Seattle area. (See locator map, left.) The black circle shows the area within a radius of 100 km of Mt. Rainier.
- Using the Select Elements tool, click on the circle to select it. Small black "handles" or a dashed line will appear around the circle when it is selected.
- Click the Select by Graphics button to select the cities within 100 km of Mt. Rainier. They will be highlighted.
- Click the Statistics button.
- In the Statistics window, calculate statistics for **only selected features** of the **Cities** layer, using the **Total Population** field.

Finding Seattle

Seattle, Washington is located near Puget Sound, on the U.S. west coast.

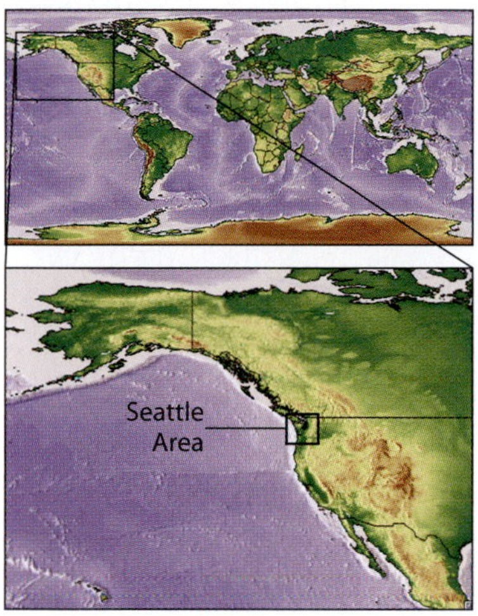

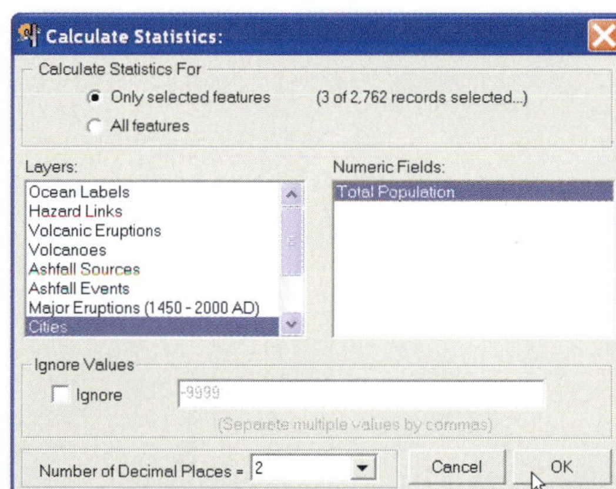

- Click **OK**. Be patient while the statistics are calculated.

The number of people affected by an eruption of Mt. Rainier is given as the **Total**.

9. How many people living within 100 km of Mt. Rainier would be significantly affected if it erupted like Mt. St. Helens?

- Close the Statistics window.
- Click the Clear Selected Features button.

Finding Mt. Pinatubo

Mt. Pinatubo is located on the island of Luzon in the Philippines.

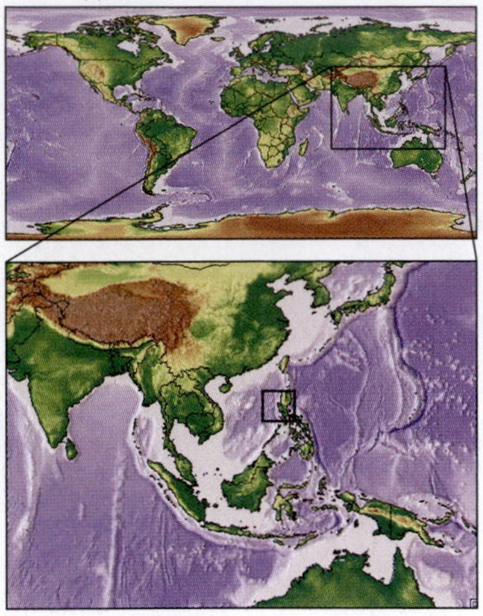

In 1991 Mt. Pinatubo in the Philippines erupted, causing widespread damage. The eruption was similar to the Mt. St. Helens eruption of 1980, only larger. With a VEI of 6, this larger blast affected an area around the volcano within a radius of about 500 km.

- Click the Full extent button to view the entire map.
- Use the locator map at the left to zoom in on Mt. Pinatubo in the Philippines. Repeat the Select by Graphics and Statistics procedures above to determine how many people would be affected by another eruption of Mt. Pinatubo.

10. How many people would be significantly affected by another eruption of Mt. Pinatubo?

- Close the Statistics window.
- Click the Clear Selected Features button.
- Quit ArcMap and do not save changes.

Reading 4.3

Volcanic hazards

Volcanoes are capable of affecting much larger areas and numbers of people than are earthquakes. Fortunately, they often provide warning signs of upcoming eruptions, allowing people to evacuate to safety. Still, there have been 35 volcanic eruptions in the past 500 years in which 300 or more people were killed.

Volcanic hazards fall into two categories.

- Direct effects—eruption clouds, shock waves from the eruption blast, lava and pyroclastic flows, and volcanic gases.
- Indirect effects—lahars (mud flows of volcanic ash), flooding, tsunamis, and post-eruption starvation.

Eruption clouds

Explosive eruptions blast rock fragments, or **tephra**, and superheated gases into the air with tremendous force. The distance these fragments travel depends on their size—smaller particles go farther than larger fragments. Volcanic **bombs**, the largest fragments, generally land within 3 km of the vent. Particles less than 2 mm (0.08 in) across, called volcanic **ash**, can rise high into the air forming a huge, billowing eruption cloud or column.

Eruption columns grow rapidly and can reach heights of more than 20 km (12 mi) in less than 30 minutes. Large eruption clouds extend hundreds of miles downwind, resulting in ashfall over broad areas. Ash from the 1980 eruption of Mt. St. Helens covered more than 50,000 km^2 (20,000 mi^2). Heavy ashfall can damage or collapse buildings. Even minor ashfalls can damage crops, electronics, and machinery.

E. Endo, USGS

A heavy load of ash from the September 1994 eruption of the Rabaul Caldera in the Philippines damaged the roof of this house. Dry ash weighs 400 – 700 kg/m^3 (880 – 1545 lb/yd^3). If ash becomes saturated with rainwater, its weight can double. In addition to the collapse risk, people trying to remove the ash face health and safety hazards.

Michael P. Doukas, USGS/CVO

Ash column of the July 22, 1980 eruption of Mt. St. Helens.

Volcanic ash clouds also pose a serious hazard to air travel. Each year, thousands of aircraft fly routes over volcanically active areas. During the past 15 years, about 80 commercial jets have been damaged by accidentally flying into ash clouds. Several have nearly crashed when ash clogged the engines and caused them to fail. The drifting cloud can carry the hazard far from its source. In 1992, Chicago's O'Hare airport was forced to close for several hours due to a drifting ash cloud from Alaska's Mt. Spurr volcano, which is nearly 5000 km (3000 mi) from Chicago!

Volcanic gases

At depth and under high pressure, magma can hold large amounts of dissolved gases. As the magma rises toward the surface and the pressure decreases, the gases bubble out and escape. This is similar to what happens when you open a can of soda.

Near the surface, the magma interacts with the surrounding rocks and ground- or surface water to generate more gases. These gases may be released suddenly in explosions or seep slowly to the surface through cracks in the overlying rock.

The behavior of magma is related to the amount of silica and water it contains. Magmas with a high silica content are generally more explosive.

Location	Source	% Silica	Explosivity
hot spots	lower mantle	50%	low
divergent boundaries	upper mantle	50%	low
convergent boundaries	upper mantle and crust	60-70%	medium to high

The table below shows magma temperatures and volcanic gas compositions (by percent) from three types of volcanic sources.

Gas	Kilauea *hot spot* *1170 °C*	Erta'Ale *divergent plate* *1130 °C*	Momotombo *convergent plate* *820 °C*
Water vapor (H_2O)	37.1	77.2	97.1
Carbon dioxide (CO_2)	48.9	11.3	1.44
Sulfur dioxide (SO_2)	11.8	8.34	0.50
Hydrogen (H_2)	0.49	1.39	0.70
Carbon monoxide (CO)	1.51	0.44	0.01
Hydrogen sulfide (H_2S)	0.04	0.68	0.23
Hydrogen chloride (HCl)	0.08	0.42	2.89
Hydrogen fluoride (HF)	-	-	0.26

As you can see, much of the gas released by volcanoes is water vapor. Some water is dissolved in the magma itself, but most water vapor is produced when the magma comes in contact with groundwater.

Carbon dioxide

Carbon dioxide is denser than air, causing it to accumulate in low areas. High concentrations of CO_2 (above 10%) are poisonous, but even lower concentrations can be deadly because CO_2 displaces oxygen required for respiration.

At California's Mammoth Mountain, the accumulation of CO_2 in the soil and in low areas is killing trees and small animals and poses a threat to humans unaware of the danger.

Lake Nyos, in Cameroon, west-central Africa, sits in a large volcanic **caldera**. A caldera is a large depression at the summit of a volcano; formed when magma erupts or withdraws from a subsurface reservoir, causing the

K. McGee, USGS

Trees killed by CO_2 near Mammoth Mountain, California.

overlying rock to collapse. Groundwater carries large amounts of dissolved CO_2 from the underlying magma into the bottom of the lake. Over time, the water becomes over-saturated with CO_2. If a disturbance causes some of this water to rise, the decreased pressure allows bubbles of CO_2 to form. The rising bubbles further disturb the water, triggering a chain reaction, resulting in a sudden, massive release of CO_2.

In 1986 a tremendous amount of CO_2 suddenly escaped from the depths of the lake, formed a cloud, and flowed down a valley toward a nearby village. Before the dense gas dissipated, it had killed more than 1700 people as well as livestock up to 25 km (16 mi) from the lake. This is not a unique event—similar cases have occurred elsewhere and will continue to occur.

J. Lockwood, USGS

Lake Nyos in western Cameroon, ten days after releasing a cloud of carbon dioxide gas that killed 1700 people in the valley below, at distances up to 25 km from the lake.

Air pollution—vog and laze

Constant eruptions, like those that occur at Hawaii's Kilauea volcano, can be significant sources of local air pollution. Under the right conditions, erupting sulfur dioxide (SO_2) gas combines with water vapor to form tiny sulfuric acid droplets. These interact with other volcanic gases, dust, and sunlight to produce volcanic smog, or **vog**. Vog reduces visibility, causes health problems, damages crops, and corrodes metals.

When molten lava comes in contact with seawater, the intense heat produces clouds of lava haze, or **laze**, which is more complex than simple clouds of steam. Heat breaks down the salt to form hydrochloric acid (HCl). The resulting cloud has a pH of 1.5 – 2.5 and can create significant health problems for downwind populations.

Pyroclastic flows

Pyroclastic flows—sometimes called **nuées ardentes** (French for "fiery clouds")—are hot, often glowing mixtures of volcanic fragments and gases that flow down the slopes of volcanoes. These flows can reach temperatures of 800 °C (1500 °F) and move at speeds up to 725 km/hr (450 mph), flattening and burning everything in their path.

The eruption of Mt. St. Helens on May 18, 1980 produced a sideways "lateral blast" that destroyed an area of 600 sq km (230 sq mi). Trees were mowed down like blades of grass as far as 24 km (15 mi) from the volcano.

C. G. Newhall, USGS

Pyroclastic flows descend the southeastern flank of Mayon Volcano, Philippines on September 23, 1984.

Lava flows and domes

When magma emerges at the surface, it is called **lava**. Lava flows bury land and structures, start fires, and may block roads or streams. It also creates new land.

How easily lava flows depends largely on its silica (silicon dioxide, SiO_2) content. Low-silica basaltic lava can form fast-moving (15 – 50 km/hr, 10 – 30 mph) streams or spread out in thin sheets many kilometers wide. **Basalt** is a dark grey or black rock formed during magmatic or volcanic activity. Since 1983, basaltic lava flows from Hawaii's Kilauea Volcano have destroyed more than 200

T.N. Mattox, USGS

Lava from a 1993 eruption of Kilauea reacts with seawater to produce plumes of highly acidic laze (lava haze).

buildings and cut off a nearby highway.

In contrast, flows of higher-silica lava tend to be thick and sluggish, traveling only short distances from the vent. These thicker lavas often squeeze out of a vent to form irregular mounds called lava domes. Between 1980 and 1986, a lava dome at Mt. St. Helens grew to a height of about 300 meters (1000 feet). Lava domes seldom pose risks to humans.

USGS

Lava flowing from Kilauea toward the sea blocks Hawaii's Highway 130.

Landslides

Volcanic landslides range from small surface movements of loose debris to collapses of the entire summit or sides of a volcano. These landslides are triggered when eruptions, heavy rainfall, or large earthquakes cause material to break free and move downhill.

Landslides are particularly common on **stratovolcanoes**, which are steep volcanoes often built up of layers of loose volcanic rock fragments. Heavy precipitation or groundwater can turn volcanic rocks into soft, slippery clay minerals, reducing the energy needed to trigger a slide.

The largest volcanic landslide in historical time occurred at the start of the May 18, 1980 Mt. St. Helens eruption. Over 2.3 km^3 (0.55 mi^3) of the mountain were transported as far as 24 km (15 mi) downhill at speeds over 240 km/hr (150 mph).

The slide left behind a **hummocky** (lumpy) deposit with an average thickness of 45 m (150 ft). For years, geologists were puzzled by a similar strange, hummocky landscape north of Mt. Shasta in northern California (shown below).

Inset photo by S. R. Brantley, USGS

Ancient landslide from Mt. Shasta, California, shown in red, was 20 times the size of the 1980 Mt. St. Helens landslide.

After witnessing the eruption of Mt. St. Helens, with its tremendous landslide and the hummocky terrain it left behind, they could then make sense of Mt. Shasta. A massive landslide occurred there about 350,000 years ago when nearly the entire volcano collapsed. Debris traveled almost 50 km (30 mi) from the volcano in an enormous landslide nearly 20 times larger than the Mt. St. Helens landslide.

Lahars

Lahar is an Indonesian term for a mixture of water and rock fragments flowing down a volcano or river valley. These flows can rush down valleys and stream channels at speeds of 30 – 60 km per hour (18 – 37 mph) and can travel more than 60 km (37 mi). As lahars pick up debris and water, they can easily grow to more than 10 times their initial size.

R. J. Janda, USGS

Aerial view of the Armero, Colombia devastation. The pattern of the city streets and buildings is still visible through the debris. The river valley from which the lahar emerged can be seen in the background.

Some lahars contain so much rock debris (60 – 90 percent by weight) that they look like fast-moving rivers of wet concrete. Close to their source, these flows are powerful enough to rip up and carry trees, houses, and huge boulders miles downstream. As slopes level out and the flows lose energy, they entomb everything in their path in mud.

Historically, lahars have been one of the deadliest volcanic hazards. In 1985, Colombia's Nevado del Ruiz volcano produced a relatively modest, VEI 3 eruption. The pyroclastic flow melted about 2.5 km^2 (1 mi^2) of snow and glacial ice on the mountain. Meltwater mixed with volcanic ash rushed down river valleys at speeds up to 60 km/hr (37 mph), stripping away rocks, soil, and vegetation.

Two hours later, a 5-m (16-ft) wall of hot mud and debris slammed into the town of Armero, Colombia, 74 km (46 mi) from the volcano, killing nearly 23,000 people and destroying 5000 homes. Surprisingly, Armero had previously been devastated by lahars in 1595 and again in 1845, only to be rebuilt on the same spot.

Volcanic resources

These short-term hazards of volcanoes are balanced by many long-term benefits to humanity. In many places, fertile volcanic soils make it profitable to live near volcanoes. Volcanic materials are used in construction and other industries. Heat energy from young volcanic systems may be harnessed to produce electricity. Most volcanoes are located near oceans, and many are found in mild climates. Because these volcanoes can lie dormant for long periods of time, people often consider them harmless.

The challenge to volcanologists is to minimize the risks associated with volcanic hazards, so that society

P.T. Pringle, Washington Division of Geology and Earth Resources

Volcanic soils are rich in minerals and have high water-holding capacities. In the state of Washington, volcanic soils around the Cascade mountains, such as Mt. Rainier, provide rich cropland.

may continue to enjoy volcanism's long-term benefits. They must continually improve their ability to predict eruptions and provide sound information to decision makers and the public.

Climate change

Major eruptions like the June 15, 1991 eruption of Mt. Pinatubo inject huge amounts of sulfur dioxide gas (SO_2) into the stratosphere. There, SO_2 combines with water to form tiny droplets of sulfuric acid (H_2SO_4) as in the diagram (below). These droplets, called **aerosols**, scatter sunlight, lowering Earth's average surface temperature for long periods of time. The satellite images (next column) show the spread of aerosols from the Pinatubo eruption.

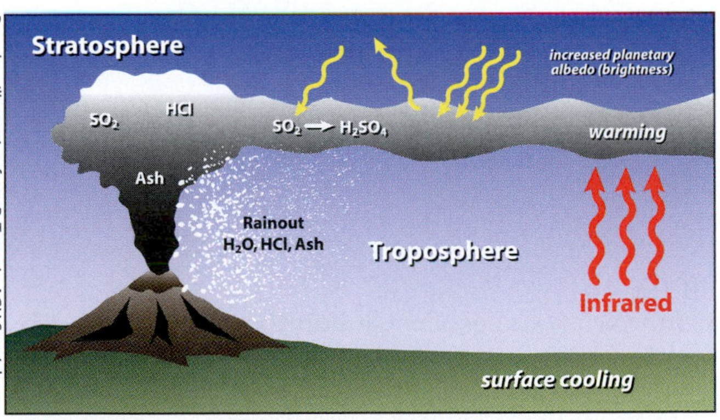

Based on illustration from R. Turco in AGU Special Report: *Volcanism and Climate Change*, May 1992.

Large eruptions pump fine ash and gases into the stratosphere. There, chemical reactions produce sulfuric acid (H_2SO_4) and other chemicals that increase Earth's albedo, or reflectivity. Less sunlight reaches the ground, causing Earth's surface to get cooler.

For two years after the Pinatubo eruption, global temperatures were about 0.6 °C (1.1 °F) lower than normal. Sulfuric acid also contributes to the destruction of the ozone layer.

These aerosols may persist for months or years, until they eventually settle out of the atmosphere or are "scrubbed" out by precipitation or other processes. Historically, periods of rapid global cooling have occurred after the largest eruptions.

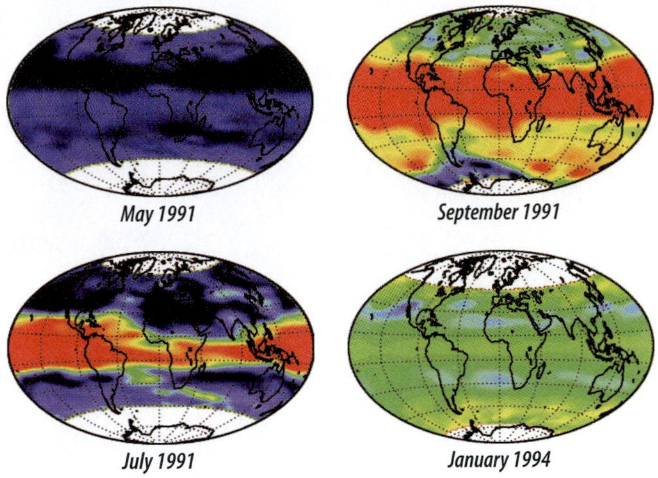

NOAA

These images show how the transparency of the atmosphere changed following the eruption of Mt. Pinatubo in June 1991. The dark blue colors of the May 1991 image indicate a clear atmosphere. After the eruption, the red band around the equator (visible in the July and September 1991 images) indicates a murky atmosphere that gradually thins as it spreads to higher latitudes (January 1994).

Questions

1. Why do volcanoes at hot spots erupt less violently than volcanoes near subduction zones?

2. What is tephra, and how does it cause damage?

3. Why is carbon dioxide gas (CO_2) a dangerous eruptive product?

4. In 1985, a relatively mild, VEI = 3 eruption of Colombia's Nevado del Ruiz volcano destroyed the town of Armero, nearly 80 km (50 mi) away. What eruptive product or process destroyed the town, and what warnings did the inhabitants have that this type of event was possible?

5. Describe three ways in which volcanoes benefit humans.

6. How does sulfur dioxide gas (SO_2) from major volcanic eruptions cause global cooling of Earth's climate?

7. How long does the cooling effect from SO_2 typically last?

Investigation 4.4

Volcanoes and climate

In this part of your investigation of volcanoes, you will look at some of the major eruptions occurring since A.D. 1450 and explore the relationship between these eruptions and changes in Earth's climate.

- Launch ArcMap, and locate and open the **etde_unit_4.mxd** file.

Refer to the tear-out Quick Reference Sheet located in the introduction to this module for GIS definitions and instructions on how to perform tasks.

- In the Table of Contents, right-click the **Volcano Hazards** data frame and choose Activate.
- Expand the **Volcano Hazards** data frame.
- Turn off all layers except **Topography** and **Countries**.

Major Eruptions and the Volcano Explosivity Index (VEI)

How often do major volcanic eruptions occur? You will look at the historical record over the last several hundred years to answer this question.

- Turn on and select the **Major Eruptions (1450 – 2000 AD)** layer.
- Select the **Major Eruptions (1450 – 2000 AD)** layer.

This layer shows the largest volcanic eruptions occurring since A.D. 1450. To determine the number of major eruptions of VEI category 5:

- Click the Select By Attributes button.
- To determine the number of VEI 5 eruptions, query the **Major Eruptions (1450 – 2000 AD)** layer for **("VEI" = 5)** as shown in steps 1 – 6. Your query will actually read:

("MAJVEI" = 5)

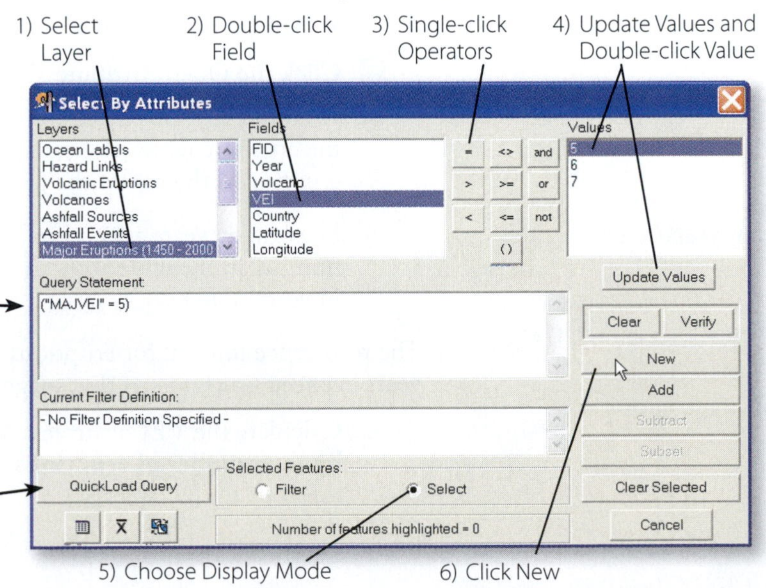

Read query statement here as you enter it.

QuickLoad Query

- Click the QuickLoad Query button and select the **Major Eruptions VEI = 5** query.
- Click **OK**.
- Click **New**.

- If you have difficulty entering the query statement correctly, refer to the **QuickLoad Query** described at left on the previous page.
- Close the Select By Attributes window.
- Click the Open Attribute Table button.
- Read the number of major eruptions at the bottom of the table (Your answer will be different than the example shown below.)

1. Record the number of eruptions with a VEI of 5 in the table below.

VEI Value	5	6	7
Number of eruptions of this VEI			
Number of years covered in data set			
Recurrence interval (average years between eruptions)			

- Close the attribute table.
- Repeat the query process above for VEI = 6 and VEI = 7, and record the number of eruptions of each magnitude in the table.
- If you have difficulty entering any of the query statements correctly, refer to the **QuickLoad Queries** described at left.
- Close the Select By Attributes window when you are finished.
- Click the Open Attribute Table button.
- Read the number of major eruptions for each query at the bottom of the table.

QuickLoad Queries

To select VEI 6 eruptions:

- Click the QuickLoad Query button and select the **Major Eruptions VEI = 6** query.
- Select the **Highlight** option.
- Click **New**.

To select VEI 7 eruptions:

- Click the QuickLoad Query button and select the **Major Eruptions VEI = 7** query.
- Select the **Highlight** option.
- Click **New**.

2. Record the number of VEI 6 and VEI 7 eruptions in the table.

- Close the attribute table.
- Click the Clear Selected Features button.

Next you will determine how many years the data set covers.

- Click the Open Attribute Table button.
- Examine the years listed for each major eruption in the **Year** column of the attribute table. They are in chronological order, starting with the oldest eruption at the top of the table.

How many years?

To find the number of years covered, subtract the year of the most ancient eruption from the year of the most recent eruption in the data set.

3. How many years are covered by the **Major Eruption** data set? Record this number in the table above. (The number is the same for all three columns in your table.)

The recurrence interval for eruptions of a particular VEI is the average number of years between eruptions of that magnitude.

4. Calculate the VEI recurrence interval by **dividing the number of years by the number of eruptions** in that VEI category. Record the recurrence intervals in the table.

5. What problem might there be with the recurrence interval you calculated for the largest VEI categories? Does it represent a full recurrence interval?

6. As the VEI increases, what happens to the number of eruptions and the recurrence interval?

Close the attribute table.

Major eruptions and Northern Hemisphere climate

In this section, you will look at a graph recently published in the science journal *Nature* that shows changes in atmospheric temperature in the Northern Hemisphere over the last 600 years.

Scientists cannot measure past temperatures directly, but they can infer temperatures indirectly by measuring the spacing of growth rings in trees. Closely spaced rings indicate poor growing conditions such as cold weather or low rainfall, and widely spaced rings indicate warmer temperatures and more rainfall.

The scientists were looking for large *anomalies,* or irregularities, in the data, which reflect years when the temperature was much lower than normal. You will examine the graph to identify these major atmospheric cooling events, then use ArcMap to determine which volcanic eruption triggered each event.

Click the Media Viewer button and choose **Volcanoes and Climate** from the list.

The graph shows how the average temperature varied from normal over a 600-year period. The bottom portion of the graph shows the VEI intensity of volcanic eruptions over the same time period. Six temperature anomalies — major atmospheric cooling events — are marked with yellow dots. They are the same as those in the following table. (Note: If needed, use the Zoom In tool to examine the graph in more detail.)

7. For each event, read the temperature anomaly on the right-hand axis of the graph and record it in the table below.

Year of anomaly	Temperature anomaly (°C)	Name of volcano	Year of eruption	VEI of eruption
1453				
1601				
1816-1818				
1884				
1912				
1992				

Close the Media Viewer window.

Tree rings record climate

Trees grow by adding new layers of woody cells just beneath the bark. Each spring in temperate latitudes, as temperatures warm up, trees begin to grow more rapidly. The cells are larger and this new wood appears lighter in color. As temperatures drop in the fall and growth slows, the cells become smaller and the wood appears darker. This pattern repeats itself yearly. By counting the number of these cycles, or rings, you can determine the age of the tree.

The width of each ring is determined by rainfall and other factors. The pattern of ring widths provides a record of the climate during the life of the tree. Matching these patterns between living and dead trees allows scientists to reconstruct long histories of climate for a region.

- Click the Open Attribute Table button.
- Scroll down the table and look for volcanoes with high VEI values (VEI > 4) that erupted in the same year as (or the year before) each temperature anomaly you recorded.

8. Use the **Major Eruptions (1450 – 2000 AD)** attribute table to complete the table. Fill in the name, date of eruption, and VEI for each eruption.
9. How well do the years of the temperature anomalies match the dates of the corresponding eruptions? What might account for differences between the two dates?

10. According to the graph, how is the VEI of an eruption related to the degree of atmospheric cooling following the event, as indicated by the temperature anomaly? Give examples to support your answer.

- Close the attribute table.

The big ones—prehistoric VEI 7 & 8 eruptions

You've seen that the 1815 Tambora eruption was a big one. The only VEI 7 eruption of the last 550 years, Tambora cooled the Northern Hemisphere by an average of 0.5 °C for several years. In fact, 1816 has been called "the year without a summer." The Northeastern U.S. was hit particularly hard, with snow in June and killing frosts from June through September. Beginning in 1817, large numbers of people started leaving northern New England for warmer climates.

The most recent powerful eruption was the VEI 6 eruption of Mt. Pinatubo in the Philippines in 1991. The eruption produced a global cooling of around 0.3 °C. In terms of total material ejected, Tambora was nearly 20 times greater than Pinatubo and about 500 times greater than the 1980 Mt. St. Helens eruption.

- Turn off the **Major Eruptions (1450 – 2000 AD)** layer.
- Turn on the **Ashfall Events** layer.
- Select the **Ashfall Events** layer.
- Turn on the **Ashfall Sources** layer.

In the not so distant past, geologically speaking, there have been tremendous VEI 8 eruptions that dwarf those of Tambora and Pinatubo. What exactly is the difference between an eruption with a VEI of 7 and one with a VEI of 8? In this next section you will compare some of the larger eruptions in human history with some of the biggest eruptions in *geologic* history by comparing the areas and volumes of their ashfall deposits.

The **Ashfall Events** layer shows the distribution of ash from several historic and prehistoric volcanic eruptions. Notice how the ash plumes overlap, particularly in what is now the United States.

- Click the QuickLoad button.
- Select **Spatial Bookmarks**, choose **Eastern Hemisphere Plumes** from the list, and click **OK**.

The volcanoes that produced the plumes are labeled by name in the figure at left. To select the ashfall plumes for Vesuvius and Pinatubo, you will probably have to zoom in even closer.

Eastern Hemisphere ashfall deposits

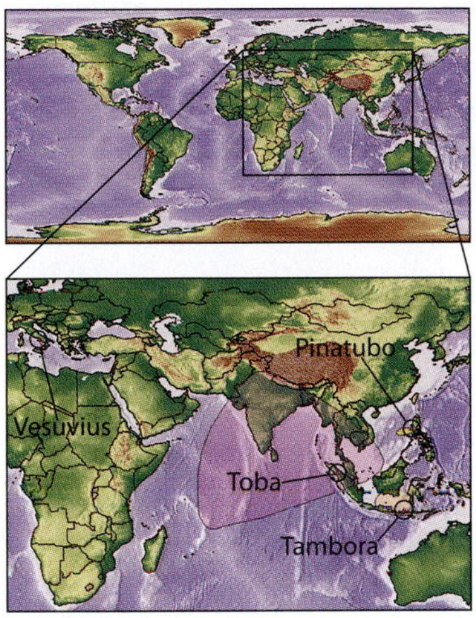

- Click the Open Attribute Table button.

11. Find the VEI of each eruption and the area of the plume it produced, and record them in the table below.

- Close the attribute table.
- Using the Measure tool, click on the source (the purple triangle), drag across to the farthest extent of each plume, and double-click to find the distance. The distance in kilometers is displayed in the lower-left corner of the map.

12. Record the distance in the table below. (Round it to the nearest 100 km.)

Volcano name	Eruption VEI	Area of ashfall plume *km²*	Maximum distance from volcano *km*
data source:	*attribute table*	*attribute table*	*measure*
Vesuvius			
Pinatubo			
Tambora			
Toba			

Now, consider the Indonesian volcano Toba, which erupted about 74,000 years ago. Toba deposited ash over an estimated 1 percent of Earth's surface, left telltale chemical traces in the ice caps of Greenland and Antarctica, and produced a prolonged period of cold and darkness that severely stressed humankind's ancestors. Genetic evidence suggests that the human population may have declined significantly due to Toba's effects, and that our species survived only by finding refuge in isolated pockets of tropical warmth. Scientists estimate that the Toba eruption caused global cooling of 3 – 5 °C (5 – 9 °F) for a period of up to seven years, with summertime temperatures in some areas dropping by as much as 15 °C (27 °F).

13. Consider the effects that the 0.5 °C cooling from the Tambora eruption had on society. How might an eruption similar in magnitude to Toba affect society today?

On a human time scale, VEI 8 eruptions are rare, occurring once or twice every 100,000 years. Geologically speaking, however, VEI 8 eruptions are relatively common. Next, you will look at three of the biggest eruptions to occur in the United States over the past 2 million years, the Yellowstone Caldera eruptions.

Caldera — a volcanic crater many times broader than a vent; formed by the collapse of the central part of a volcano.

Crater — circular depression (smaller than a caldera) created by an explosive excavation of rock during eruptions.

The Yellowstone Caldera eruptions

Occupying the northwest corner of Wyoming is a spectacular area of geysers, hot springs, and abundant wildlife. This region was so unusual and notable for its geological and biological wonders that it was protected back in 1869 as the United States' first national park. Called Yellowstone, the park today is considered one of the three "crown jewels" of the National Park System, and receives about 3 million visitors per year.

Yellowstone Caldera ashfall deposits

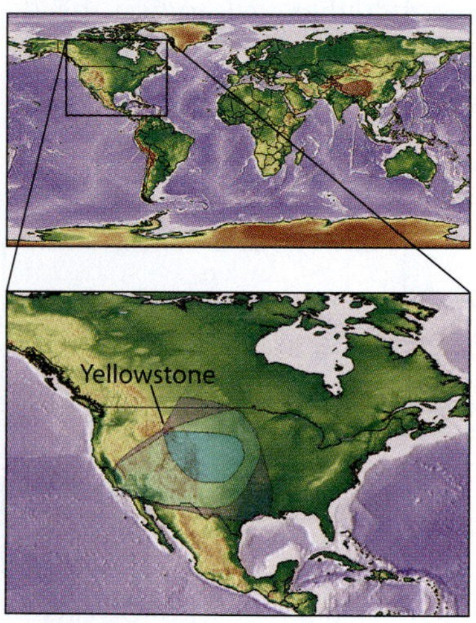

Many visitors are unaware that Yellowstone is a giant collapsed volcano, or caldera, that produced three gargantuan eruptions within the past 2 million years. Next, you will take a closer look at this slumbering volcanic giant.

- Click the Full extent button to view the entire map.
- Use the Zoom In tool to zoom in on the continental United States.

You should be able to see several overlapping ashfall plumes across the middle and southwestern parts of the U.S. Three of these plumes were produced by volcanic activity at Yellowstone; the fourth plume was created by another massive eruption from the Long Valley Caldera in California.

To highlight the Yellowstone ashfall plumes:

- Click the Select By Attributes button.
- Click the **QuickLoad Query** button inside the Select By Attributes window.
- Select the **Yellowstone Caldera Deposits** query and click **OK**.
- Click the **New** button.
- Close the Select By Attributes window.

Now the Yellowstone ashfall plumes should be outlined on your map. Next you will use the **Ashfall Events** attribute table to obtain information about each eruption.

- Click the Open Attribute Table button.

14. Complete the following table with information about the three Yellowstone ashfall deposits. List them in order from oldest to youngest.

Name of deposit	Age *years before present*	Area of deposit *km²*

15. Estimate the area in square kilometers that the ashfall from the next Yellowstone Caldera eruption might cover. You can do this by calculating the average area of the three historical ashfall deposits.

Close the attribute table when you are finished.

Finally, calculate the recurrence interval. You can use the recurrence interval to predict when the next major eruption from the Yellowstone Caldera might occur.

- Subtract the youngest (most recent) age from the oldest age in the table above.
- Divide by the number of times the original eruption was repeated; in this case, divide by 2.

16. What is the recurrence interval for the Yellowstone Caldera?

17. Use the most recent eruption and the recurrence interval to estimate when the next eruption might take place.

Quit ArcMap and do not save changes.

Unit 5
Tsunami Hazards

In this unit, you will

- ***Analyze two major tsunami events in detail.***
- ***Discover the effects tsunamis have on communities and how communities can prepare for them.***
- ***Examine tsunami trigger events and develop criteria for issuing tsunami warnings.***

NOAA/NGDC/Sunset Newspaper

A major earthquake off the coast of Chile on May 22, 1960 produced a tsunami that affected the entire Pacific Basin. In Hilo, Hawaii — 10,000 km from the earthquake — the tsunami caused 61 deaths and $24 million in property damage. Frame buildings were either crushed or floated off their foundations, and only buildings of reinforced concrete or structural steel remained standing.

Warm-up 5.1

Scotch Cap Lightstation

Location map

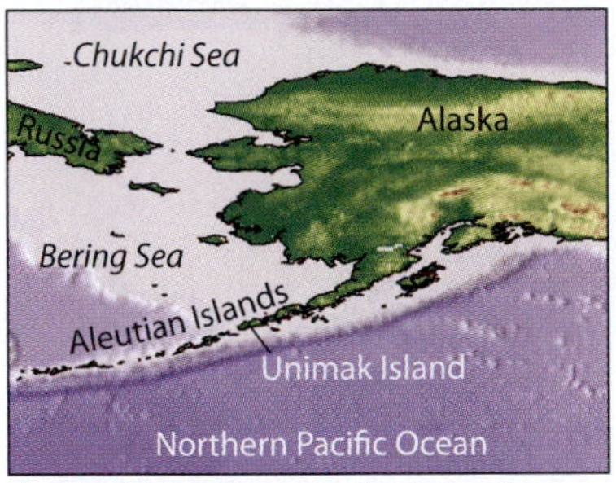

The Scotch Cap Lightstation, located on Unimak Island in Alaska, was a five-story, reinforced concrete structure built 10 meters (33 feet) above sea level. On April 1, 1946 a strong earthquake jolted the lightstation. Forty-five minutes later, a tsunami estimated at 35 meters (115 feet) high obliterated the lightstation.

The eyewitness account reprinted here is the actual report submitted by a petty officer stationed in the radio shack seen on the cliff above the lightstation in the bottom picture.

After reading the account of the tsunami that destroyed the Scotch Cap Lightstation in 1946 (pages 117–118), list and describe all of the tsunami-related hazards discussed. Feel free to add other hazards from your previous knowledge or experience with tsunamis.

Tsunami hazards

1.

2.

3.

4.

5.

6.

7.

8.

US Coast Guard

Before (above) and after (below) pictures of the Scotch Cap Lightstation.

Tsunami hazard examples (optional)

If you have access to a computer, you can see examples of these and other tsunami hazards. Be sure to add any new hazards you find to your list.

- Launch ArcMap, and locate and open the **etde_unit_5.mxd** file.

If you encounter unfamiliar GIS terminology or are unable to perform a task, refer to the Quick Reference Sheet in the introduction to this module.

- In the Table of Contents, right-click the **Geological Hazards** data frame and choose Activate.
- Expand the **Geological Hazards** data frame.

To see examples of other earthquake hazards:

- Turn on the **Hazard Links** layer.
- Select the **Hazard Links** layer.
- Using the Hyperlink tool, click on each of the blue tsunami hazard symbols (blue waves) on the map.
- Read the caption for each picture (or watch the movie), then close its window. There may be more than one picture for each link.

Questions

1. Which of the hazards you listed on the previous page caused the greatest amount of damage and loss of life at the Scotch Cap Lightstation?

2. Do you think anything could have been done to protect the lightstation or its staff? Explain.

3. From the account, did the lightstation staff have warning signs of the coming wave? Explain.

4. Why do you think so much of the world's population lives close to the ocean?

5. Would you rebuild the lightstation in the same location? Explain.

Eyewitness account

Scotch Cap Lightstation

Memorandum kept by Chief Radio Electrician
Hoban B. Sanford, U. S. Coast Guard

At 0130 Xray, 1 April, 1946, at which time I was awake and reading, a severe earthquake was felt. The building (CG Unit 368 - Unimak A/F Station) creaked and groaned loudly. Objects were shaken from my locker shelves. Duration of the quake was approximately 30 to 35 seconds. The weather was clear and calm.

Knowing that the volcanoes to Northward of the building had been active at one time, I immediately looked in that direction for signs of renewed activity and upon seeing none made a round of the building to see what, if any, damage had been caused by the tremor. Inspection failed to reveal any damage other than objects shaken from locker shelves. The crew were all awakened by the quake.

Intending to call Scotch Cap Lightstation on the phone to ascertain if they had felt, or been damaged by, the quake, I went to the phone in Operations, but Pitts, RM2c, had already done so and he stated they had felt the tremor and that Pickering, MoMM2c, who was on watch at the lightstation, had said that he was "plenty scared" and was going to call Dutch Harbor Navy Radio to see what information that unit might have regarding the earthquake.

At 0157 Xray a second severe quake was felt. This one was shorter in duration, lasting approximately 15 to 20 seconds, but harder than at 0130 Xray. I again looked towards the mountains for any signs of volcanic activity, but still could see none. I made a second round of the building to see if any damages had resulted but none was apparent.

The crew was gathered in the Recreation Hall discussing the shocks, their probable cause and location, when a crew member stated he had talked with Scotch Cap Lightstation after the second shock, and they were attempting to contact Dutch Harbor Radio for any news of the quakes.

At 0218 Xray a terrible roaring sound was heard followed almost immediately by a very heavy blow against the side of the building and about 3 inches of water appeared in the galley, Recreation Hall and passageway. From the time the noise was heard until the sea struck was a matter of seconds. I should say between five and ten seconds at most. Ordering the crew to get to the higher ground of the DP D/F building immediately. I went to the control room and, after a couple of calls to Kodiak and Adak Net Control Stations, broadcast a priority message stating we had been struck by a tidal wave and might have to abandon the station, and that I believed Scotch Cap Lightstation was lost.

> (Message: - PPP NMJ NNA NNFV NNBT TIDAL WAVE MAY HAVE TO ABANDON THIS PLACE X BELIEVE NNHX LOST INT R INT R XXX)

Received no answer to calls or receipt for message and did not know until daylight that the receiving antennae had been carried away. Electric power was fluctuating badly, and starting for the generator room to ascertain the cause and extent of damage, I found that D'Agostino, ETMlc, and Campanaro, RM2c, had voluntarily remained behind to assist.

Water had struck the switchboards through a burst-in door and the voltage control regulator was burning on the back of the board. ETMlc D'Agostino used a CO_2 extinguisher while I shut down the generator. This placed the station in darkness.

Companaro found and lit a kerosene pressure lantern and we proceeded to make emergency repairs. The switchboards were shunted and the generator connected directly to the line. This restored lighting and some power circuits. Companaro was sent to call back some members of the crew to get more clothing and canned goods to be taken to the DP building in case of a second wave. While crew members were thus engaged, D'Agostino and myself made a rapid survey of damage. At 0345 I went to edge of hill above Scotch Cap Lightstation to observe conditions there. The way was littered with debris, and the lightstation had been completely destroyed. I returned to the DF Station and, with D'Agostino, continued cleaning up water and muck about generators. At 0550 had one generator running full power, at this time transmitted a dispatch to the DCGO, 17ND, via Kodiak re conditions. At 0700 went down to the site of the lightstation, the sea by this time having receded to its usual limits, and in company with several crew members searched among the debris for any signs of bodies of personnel. On top of hill behind the lightstation we found a human foot, amputated at the ankle, some small bits of intestine which were apparently from a human being, and what seemed to be a human knee cap. Nothing else was found. At 0725 was informed Sarichef Beacon heard. At 0800 sent out searching parties to attempt locate any trace of Scotch Cap personnel. Searching parties later returned and reported no trace of the lightstation crew. The crew of the lightstation was comprised of Petit, CBM, OinC; Pickering, MoMM2c; Dykstra, Slc; Ness, Slc; and Calvin, Flc.

Searching parties were out daily whenever weather permitted until 20 April when CBM Sievers of CGC *CLOVER*, which was establishing a temporary light on the site of the destroyed light, located a body which was identified as Paul J. Ness, Slc, a member of the lightstation crew. The body was viewed by several crew members and myself and all agreed that it was Ness, who had high cheek bones, slightly prominent upper incisor teeth and a small goatee. The pharmasist [*sic*] mate from Unit 368 had been observing the large toes on both feet of Ness and the nails were pared away from the sides. This condition, also existed on the feet of the body. The remains were wrapped in an old blanket and canvas and removed to above the high water mark, pending burial instructions from DCGO, 17ND. On 22 April at 1030 CBM Sievers, who was conducting a search to eastward, returned to Unit 368 and stated he had found another body. With several crew members I proceeded on to the location but was unable to identify the body. The body was decapitated, disemboweled, and in a poor state of preservation. A homemade monel ring on the right hand could not be identified by any member of the crew of Unit 368.

At 1100 crew members who had been searching to westward reported they had found the right thigh and foot of a man. The foot could not be identified. These remains were gathered in old mail sacks and placed in a rough coffin. The body of Ness was placed in an individual coffin.

At 1545, 23 April, the body of Ness was buried in an individual grave, the unidentified portions of bodies were buried in a common grave adjacent thereto. The graves are at the seaward edge of the western bank of the first ravine to the eastward of Scotch Cap Lightstation and are approximately 300 yards from the site of the light, near the graves of two Russian seamen. The graves are plainly marked with white wooden crosses with brass plates securely attached, and are well covered with rocks to discourage depredation by animals.

The area covered by searches was approximately 5 miles eastward, 4 miles westward from Scotch Cap Lightstation, and inland to the high water mark of the tidal wave.

Notes: Most of the abbreviations used in this account are U.S. Coast Guard occupations and ranks: RM2c - Radioman Second Class • MoMM2c - Motor Machinist's Mate Second Class • ETM1c - Electronic Technician's Mate First Class • S1c - Seaman First Class • F1c - Fireman First Class • CBM - Chief Boatswain's Mate • OinC - Officer In Charge. Other abbreviations include the following: DCGO - District Coast Guard Officer • CGC - Coast Guard Cutter. "Xray" is a designation for the time zone covering Unimak Island. It is UTC −11, which means it is 11 hours offset from Greenwich Mean Time. For comparison, the U.S. Eastern time zone is UTC – 5, Central is UTC – 6, Mountain is UTC – 7, and Pacific is UTC – 8. "0130 Xray" at Unimak Island is thus 7:30 in the morning on the East Coast and 4:30 in the morning on the West Coast.

Source: U.S. Coast Guard

Investigation 5.2

Deadly tsunamis

What does it mean?

The Japanese characters for tsunami translate literally as "harbor wave." Understanding tsunami behavior is important in Japan, where both lives and property have been destroyed by tsunamis throughout history.

津 tsu = *harbor* or *port*

波 nami = *wave*

Tsunamis (soo-NAH-meez) are one of the most devastating geologic hazards facing coastal communities. Sometimes mistakenly called "tidal waves," tsunamis are not related to the ocean's tides. Rather, they are waves caused by the sudden movement of the seafloor or coastline due to geologic events such as earthquakes, volcanic eruptions, or landslides.

When the seafloor moves suddenly, it displaces the overlying water, forming a "hill" or "valley" on the surface. As gravity makes the ocean surface return to its normal level, the energy spreads out rapidly as a series of concentric waves, like ripples on a pond.

Tsunamis are not like wind-blown ocean waves. Wind-blown waves have amplitudes (height ÷ 2) up to a few meters and wavelengths (distance from crest to crest) of a few hundred meters. In the open ocean, tsunami amplitudes range from a few millimeters to about half a meter depending on the depth of the water, and the tsunamis have wavelengths of hundreds of kilometers.

Period [of a wave] — the time required for successive wave crests to pass a fixed point.

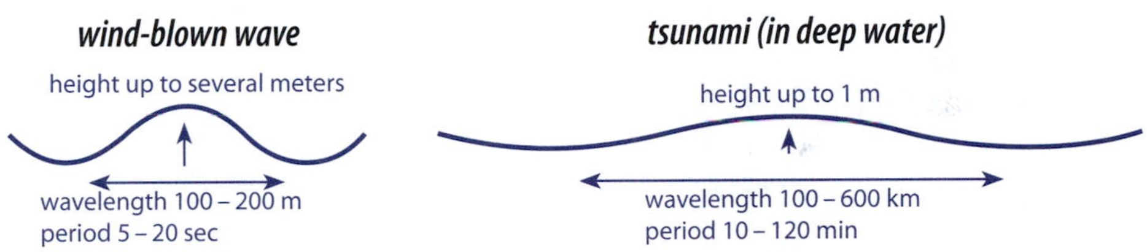

The great tsunami of 1960

Launch ArcMap, and locate and open the **etde_unit_5.mxd** file.

If you encounter unfamiliar GIS terminology or are unable to perform a task, refer to the Quick Reference Sheet in the introduction to this module.

In the Table of Contents, right-click the **1960 Chile Tsunami** data frame and choose Activate.

Expand the **1960 Chile Tsunami** data frame.

On May 22, 1960 a magnitude 9.6 earthquake, the strongest ever recorded, struck off the coast of Chile. The location of the quake is marked in this data frame with a red star symbol.

Turn on the **Plate Boundaries** layer.

1. At which type of plate boundary did the earthquake occur?

Turn off the **Plate Boundaries** layer.

The quake generated a huge tsunami that temporarily raised sea level by as much as 30 m and devastated communities along the Chilean coast. In all, more than 1200 people were killed. Unfortunately, the danger was not limited to the nearby coastline.

In the open ocean, tsunamis are difficult to detect and impossible to see. To study tsunami behavior, researchers build numerical models on computers. One of these models recreates the 1960 Chilean tsunami. The heights of the waves are greatly exaggerated to make them visible.

The animation does not show the entire globe. The star marks the location of the earthquake that produced the tsunami.

The model shows only part of the globe. Use the diagram at left to orient yourself before viewing the animation.

To view the 1960 Chilean tsunami model:

- Using the Hyperlink tool, click on the tsunami source off the coast of Chile (the red star). Be patient while the movie loads.
- Watch the movie several times. The waves travel a long distance, so the view rotates twice to show them passing Hawaii and reaching the far side of the Pacific.

After viewing the movie, answer the following questions.

2. How do tsunamis respond when they strike islands or coastlines?

Animation courtesy of Disaster Control Research Center, Tohoku University, and Obayashi Corporation, Japan.

3. By the time the waves reach Japan, how much energy do they appear to have lost? (All? Most? A little? None?)

- Close the Media Viewer window.

Tsunamis exhibit all of the properties of waves. In the animation, notice how tsunamis "slosh" back and forth in the ocean as they reflect off coastlines, islands, and shallow underwater landforms. They reflect (bounce) off obstacles, refract (change direction) as they change speed, and diffract (change direction) as they pass obstacles. As they slosh around the ocean basin, colliding waves may add together or cancel each other out, causing their amplitudes to be much higher or lower than expected.

Measuring the effects of tsunamis

Often, the first indication of a tsunami is a rapidly falling water level, followed by a rapidly rising water level that inundates low-lying coastal areas. Major tsunamis can also have towering breaker waves. A tsunami's impact on these areas is usually described by two measurements:

- **run-up** — the maximum height of a tsunami above normal sea level.
- **run-in** — the distance the rising water reaches inland from the normal coastline.

- Turn on the **Maximum Run-up** layer.

Select the **Maximum Run-up** layer.

This layer shows locations that recorded measurable run-up from the 1960 Chilean tsunami. As you saw in the animation, this tsunami affected the entire Pacific basin.

The color of the run-up symbols varies with the height of the run-up. As expected, sites near the source recorded the highest run-up, with the run-up generally decreasing with distance. Unusual circumstances, however, may produce a high run-up far from the source. Next, you will examine the effects of the tsunami on the islands of Hawaii and Japan.

Run-up and run-in

Tsunami run-up and run-in depend on many factors, including the size of the trigger event, the distance traveled, seafloor topography, tides, and the shape of the coastline.

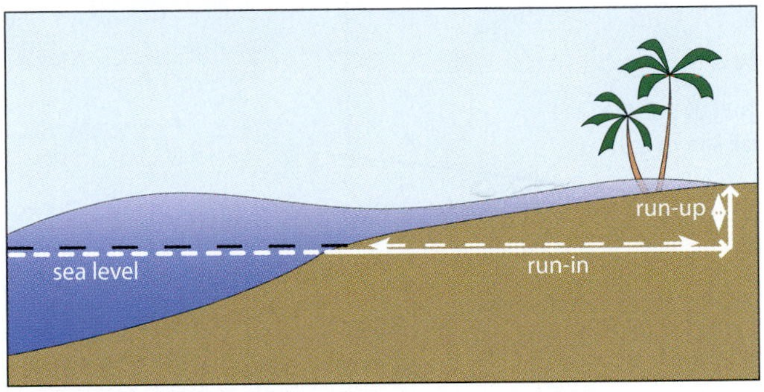

Calculating the tsunami's speed

To calculate the Chilean tsunami's speed, you need to know the distance it traveled and the time it took to go that distance.

Using the Measure tool, measure the distance from the tsunami source to Hawaii by clicking once on the source, dragging across to Hawaii, and double-clicking on Hawaii. The distance (Total) is displayed in the Status Bar in the lower left corner of the data frame window.

4. Record the distance from the trigger event to Hawaii in column 2 of the table below. The distance to Japan has already been recorded for you.

Location	Distance *km*	Shortest travel time *hours*	Speed *km/hr*	Highest run-up *m*
data source:	*measured*	*from table*	*calculated*	*from table*
Hawaii				
Japan	17,000			

Next you will find the shortest time required for the tsunami's leading wave to reach Hawaii.

Click the Select By Attributes button.

To display run-up sites in Hawaii, query the **Maximum Run-up** layer for **("Run-up Location" = 'HAWAII')** as shown in steps 1-6 on the following page. Your query will actually read:

("RUNUP_LOCA" = 'HAWAII')

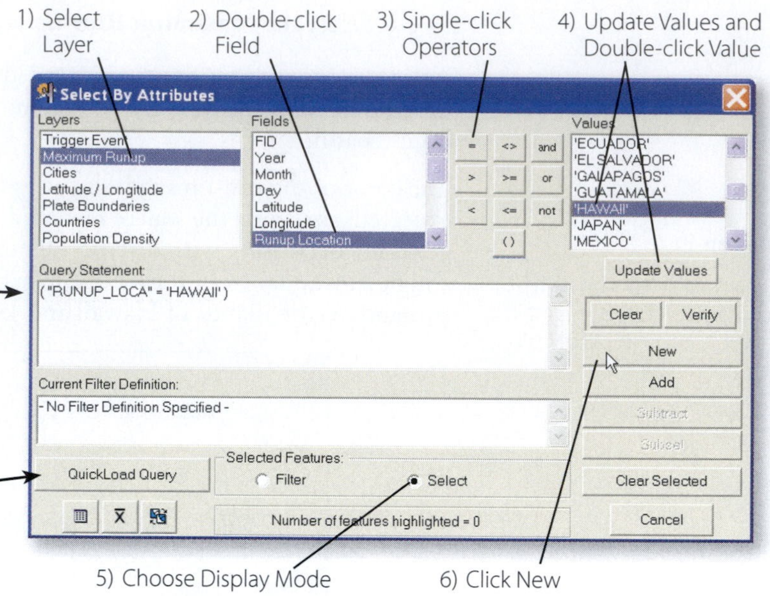

Read query statement here as you enter it.

QuickLoad Query

- Click the QuickLoad Query button and select the **Hawaii Run-up** query.
- Click **OK**.
- Click **New**.

If you have difficulty entering the query statement correctly, refer to the **QuickLoad Query** described at left.

The run-up sites in the Hawaiian Islands should now be highlighted on your map.

Click the Statistics button $\bar{X}$ in the Select By Attributes window.

In the Statistics window, calculate statistics for **only selected features** of the **Maximum Run-up** layer, using the **Travel_Time** field.

Choose the **Ignore** option and enter a value of **0**.

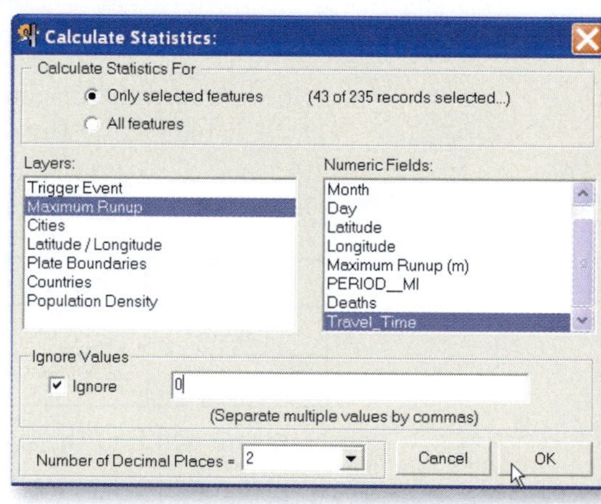

Travel-time data format

The travel-time data are given in decimal hours, not hours and minutes.

Click **OK**. Be patient while the statistics are calculated.

The shortest travel time is reported as the **Minimum**.

5. Record the shortest travel time (hours) in the table on page 121.

Close the Statistics window.

Click the Statistics button $\bar{X}$ in the Select By Attributes window.

In the Statistics window, calculate statistics for **only selected features** of the **Maximum Run-up** layer, using the **Maximum Run-up (m)** field.

- Click **OK**. Be patient while the statistics are calculated.

The highest run-up is reported as the **Maximum**.

6. Record the highest run-up (m) in the table on page 121.

- Close the Statistics window.

To find the shortest time required for the leading wave to reach Japan:

- Click the **Clear** button in the Select By Attributes window.
- To display run-up sites in Japan, query the **Maximum Run-up** layer using the query statement **("Run-up Location" = 'Japan')**. Your query will actually read:

("RUNUP_LOCA" = 'JAPAN')

- Click the **New** button.
- If you have difficulty entering the query statement correctly, refer to the **QuickLoad Query** described at left.

QuickLoad Query

- Click the QuickLoad Query button and select the **Japan Run-up** query.
- Click **OK**.
- Click **New**.

- Repeat the statistics procedure above to find and record the shortest travel time and maximum (highest) run-up for sites in Japan in the table on page 121. (Hint: Remember to choose the **Ignore** option and enter a value of **0** when calculating statistics on travel times.)

7. Calculate the average speed of the tsunami between its source and each location using the formula: speed = distance ÷ time, and record it in the table on page 121.

- Close the Statistics and Select By Attributes windows.
- Click the Clear Selected Features button.

The average depth of the Pacific Ocean between Hawaii and Japan is greater than it is between Chile and Hawaii.

8. Based on the speeds you calculated, what happens to the speed of a tsunami when the water depth increases?

9. Based on the highest run-up recorded at each location, what happens to the energy of the tsunami as it travels away from its source?

Local effects

- Click the QuickLoad button.
- Select **Data Frames**, choose **1993 Japan Tsunami** from the list, and click **OK**.

At 10:17 p.m. local time on July 12, 1993 a magnitude 7.8 earthquake struck in the Sea of Japan off the coast of Hokkaido, generating one of the worst tsunamis in Japanese history. Run-up reached as high as 30 m on the nearby island of Okushiri. (See locator map in the sidebar on the following page.) In the map on your screen, the red star shows the location of the quake's epicenter.

Location of the 1993 Japan tsunami

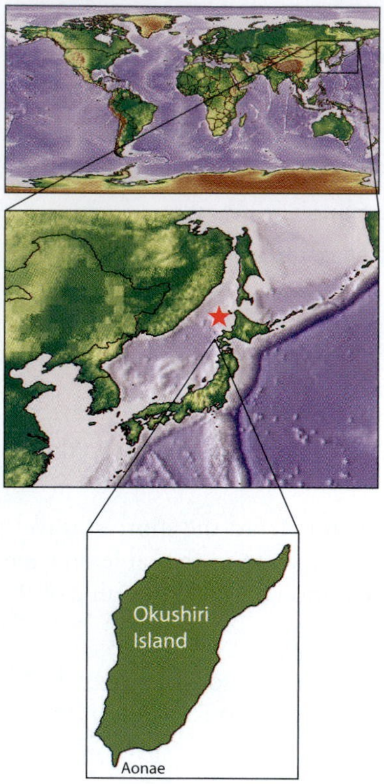

Note: Your map does not show the island in this much detail.

Particularly hard-hit was the resort town of Aonae, on the southern tip of the island, where 185 people lost their lives.

Researchers have modeled the impact of the tsunami on Aonae to help understand what happened and to learn how to better protect coastal communities from damage.

To view an animation of the effects of the tsunami on Aonae:

- Select the **Trigger Event** layer.
- Using the Hyperlink tool, click on the red star symbol. Be patient while the movie loads.
- The animation shows, from several different angles, the tsunamis striking the coastline. Watch the movie several times, then answer this question.

10. In the animation, where do the waves appear to wash the highest and farthest inland? Where do you think are the most dangerous places to live in this area?

- Close the Media Viewer window.

To see before and after pictures of damage to the Aonae peninsula:

- Click the Media Viewer button and choose **Aonae Before & After** from the media list.

The tsunami struck at night, and the "After" picture shows what the area looked like the morning after the tsunami.

- Close the Media Viewer window.

How much warning?

In the 1960 tsunami, it took many hours for the wave to reach Hawaii and Japan. Had an effective warning system been in place, many lives could have been saved. Next, you will find out how much warning the community of Aonae had. Could a warning system have saved the people who died there?

Aonae locator map

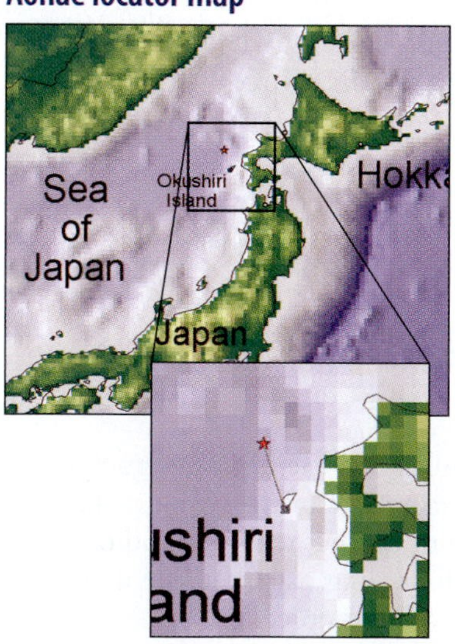

- Turn on the **Maximum Run-up** layer, then zoom in on the Sea of Japan until you can see the run-up symbol on the southern tip of the island of Okushiri. (See locator map at left.)
- Using the Measure tool, measure the distance from the tsunami trigger event to the Aonae run-up symbol.
- Read the distance from the status bar in the lower left corner of the data frame.

11. Assuming that the tsunami traveled at about 15 km/min (900 km/hr), calculate the travel time to Aonae in minutes using the formula:

$$\text{time} = \text{distance} \div \text{speed}$$

12. Do you think a tsunami warning system could have saved many lives, based on the travel time you calculated in question 10? Explain.

Quit ArcMap and do not save changes.

Reading 5.3

Analysis of a tsunami

Introduction

Life on Earth evolved in the protective environment of its oceans. To this day, humankind continues to take advantage of the benefits of living along the shore. Moderate temperatures, abundant food, and easy transportation are all provided courtesy of our planet's seas and oceans. Living near the ocean is not without its perils though. Perhaps the most unpredictable and terrifying of these hazards is the tsunami.

A tsunami is a series of large waves created when a disturbance displaces, or moves, an enormous volume of ocean water. The waves of a tsunami spread outward from their source at very high speed. This means that tsunamis can cross an entire ocean in a matter of hours, making them a truly global hazard for coastal communities.

Tsunamis are a special class of waves called **shallow-water waves**. The speed at which they travel is proportional to the depth of the water. In the open ocean, tsunamis move very fast but have wave heights of 1 m or less. As the water shallows near coastlines, the speed of the waves decreases while their height increases. When the waves reach the shore, they may be tens or, in extreme cases, hundreds of meters high.

Two measurements are used to describe the effect of tsunamis on the coastline.

- **Run-up** is the maximum height of the tsunami wave above normal tide level.
- **Run-in** is a measure of how far inland the wave reaches beyond the normal shoreline.

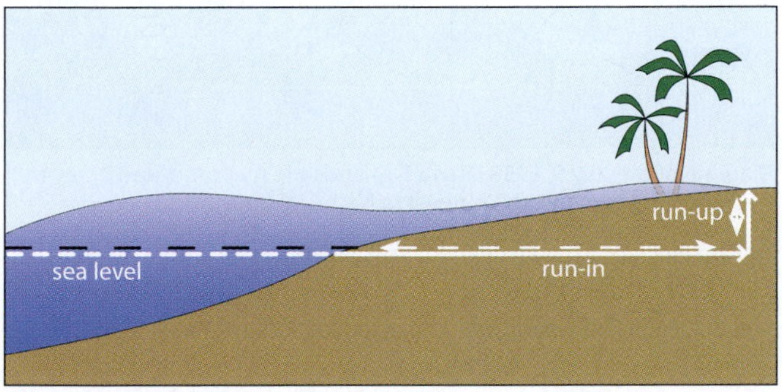

Tsunami magnitude scale

Tsunami researchers have developed a scale, based on the run-up height, for describing the intensity of a tsunami.

Intensity	Run-up height *m*	Description	Frequency in Pacific Ocean
4	16	Disastrous. Near complete destruction of man-made structures.	1 in 10 years
3	8	Very large. General flooding, heavy damage to shoreline structures.	1 in 3 years
2	4	Large. Flooding of shore, light damage to structures.	1 per year
1	2	Moderate. Flooding of gently sloping coasts, slight damage.	1 per 8 months

Tsunami magnitude (similar to earthquake intensity) is a measure of the local size of a tsunami.

Where tsunamis occur

In the Pacific Ocean, where the majority of tsunamis occur, the historical record shows extensive loss of life and property. Japan, in particular, has repeatedly seen entire towns and cities wiped out by tsunamis, most recently in 1993.

Most coastal communities have some degree of tsunami risk, but the chart shows that tsunamis are far more common in Pacific coastal areas than on the U.S. East Coast. This is because earthquakes and volcanoes, the features that most commonly cause tsunamis, occur more often in the Pacific Basin than in the Atlantic Basin.

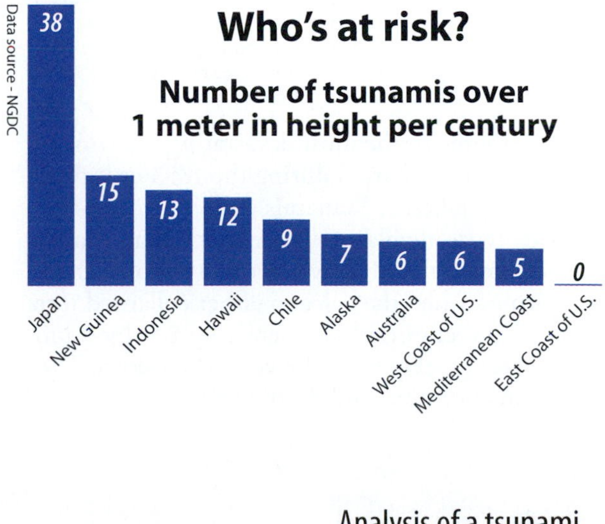

Tsunamis of all sizes

Tsunamis occur at various scales, depending on the magnitude of the event that triggers them, the location, and the surrounding topography.

- **Local tsunamis** affect an area within 200 km of their source.
- **Regional tsunamis** affect an area within about 1000 km of their source.
- **Teletsunamis** travel great distances (over 1000 km), often across entire oceans.

What causes tsunamis?

There are four types of events capable of producing tsunamis: earthquakes, volcanoes, landslides, and asteroid impacts.

Earthquakes—Normally, only large earthquakes in or near ocean basins produce tsunamis. Of these, only the strongest—magnitude 8 and higher—create teletsunamis. However, smaller earthquakes sometimes indirectly cause tsunamis by triggering landslides.

USGS

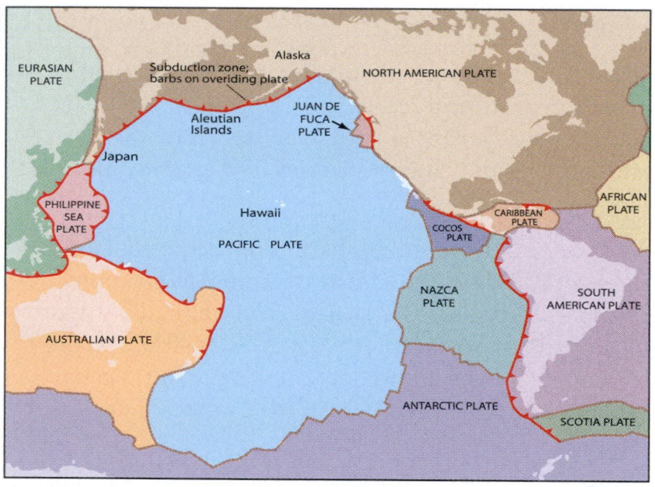

This map explains the prevalence of tsunamis in the Pacific Ocean. The red lines are subduction zones, where one tectonic plate is plunging beneath another. Earthquakes and volcanoes, two common triggers for tsunami events, are typical features of subduction zones.

Volcanoes—Some of the most devastating tsunamis in recorded history occurred during the 1883 eruption of the volcano Krakatoa. Tsunamis with 40-m run-ups destroyed 165 coastal villages on the Indonesian islands of Java and Sumatra, killing over 36,000 people. The tsunami formed when the volcano either collapsed into its magma chamber or exploded, creating a 5- by 9-km crater. Seawater quickly filled the void, then sloshed outward as an enormous and deadly tsunami.

Landsat 7/USGS

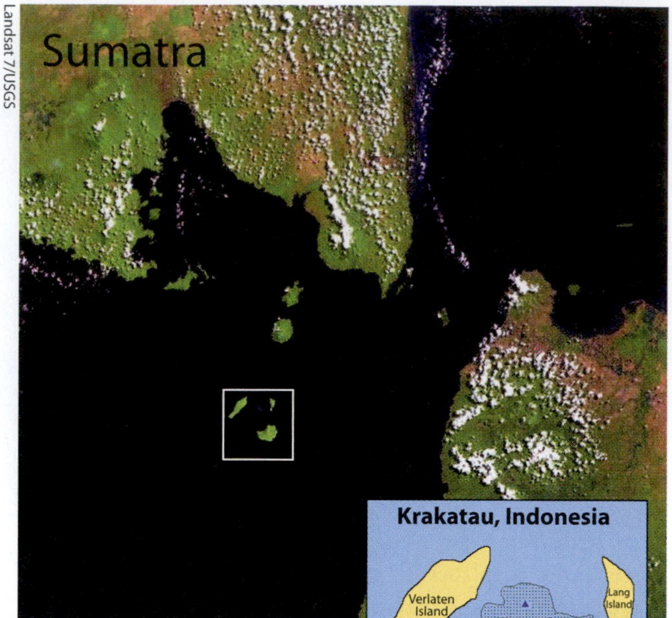

Krakatoa (Krakatau) sits in the Sunda Strait, a major shipping lane between the Indian Ocean and the Java Sea.

Lyn Topinka, USGS/CVO

Landslides—The movement of rocks and soil can displace large volumes of water, creating a tsunami. These landslides can develop on land and fall into the water or take place completely underwater. One of the highest tsunamis in modern history occurred in Lituya Bay, Alaska when a magnitude 8.3 earthquake triggered a landslide that fell into the nearly enclosed bay. The slide created a tsunami splash wave that washed 524 m (1720 ft) over a ridge on the opposite side of the bay.

Dr. Charles Mader/The Tsunami Society

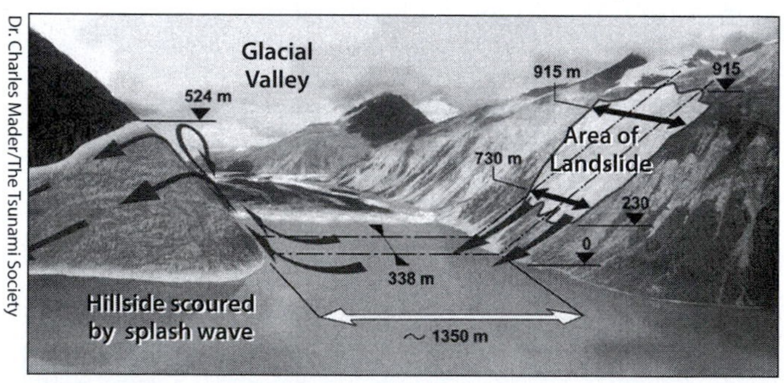

Diagram of the July 9, 1958 Lituya Bay landslide that produced a 524-m (1720-ft) local tsunami, the highest in historical times.

Scientists are also investigating submarine landslides for their tsunami-causing potential. Far more massive than terrestrial landslides, underwater landslides off the Hawaiian Islands have sent thousands of cubic kilometers of material sliding almost 200 km from their source.

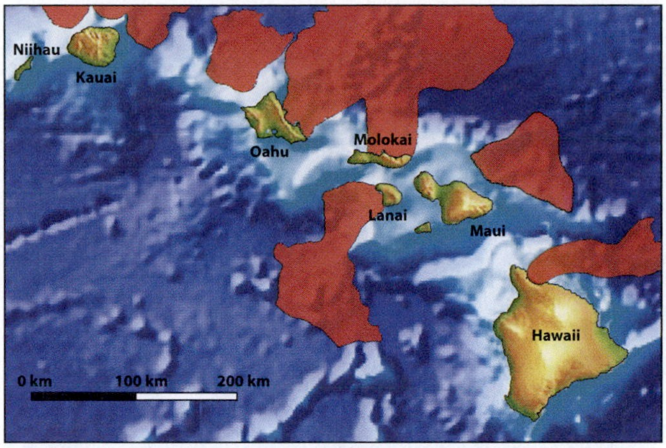

Locations of major Hawaiian submarine landslides. Similar landslides may threaten the northwestern and eastern U.S. coasts with tsunamis.

The resulting tsunamis would have been enormous. There is speculation today about the origin of coral boulders 200 m above sea level on the Hawaiian Island of Molokai. Were they tossed there by a massive tsunami, or deposited by normal processes when the sea level was significantly higher?

Asteroid [and comet] impacts—The rarest yet most catastrophic cause of tsunamis is an asteroid [or comet] impact in one of Earth's oceans. Scientists estimate that a 1-km diameter asteroid striking the middle of an ocean would produce a tsunami with run-ups ranging from 6 to 50 m.

The asteroid that hit near Mexico's Yucatán Peninsula 65 million years ago, contributing to the extinction of the dinosaurs, was about 10 km in diameter. Evidence suggests that the impact generated a tsunami 100 to 250 m high that washed hundreds of kilometers inland.

Record of a tsunami event

A tsunami is not a single wave, but a series of wave build-ups and retreats. The waves may be spaced minutes or hours apart. People often believe that the tsunami danger is over after the first wave has passed and are then killed when they return to coastal areas to clean up after the event.

To complicate matters, tsunamis reflect off coastlines and may pass a given location several times and from different directions. The highest run-up can occur several hours after the arrival of the first wave. Tides are another key factor, as run-up is more severe when it occurs at high tide than at low tide. All of these features of a tsunami event can be seen in the tide gauge record from the 1960 Chilean tsunami shown below.

Tsunami effects

Direct Impact—Most of a tsunami's initial damage comes from the direct impact of waves on structures and harbor facilities, and from wave run-up on coastal buildings.

To get a better idea of the damage a large tsunami can do, consider that a cubic meter of pure (i.e., fresh) water has a mass of 1000 kg (2200 lb). Therefore, a 15-meter-

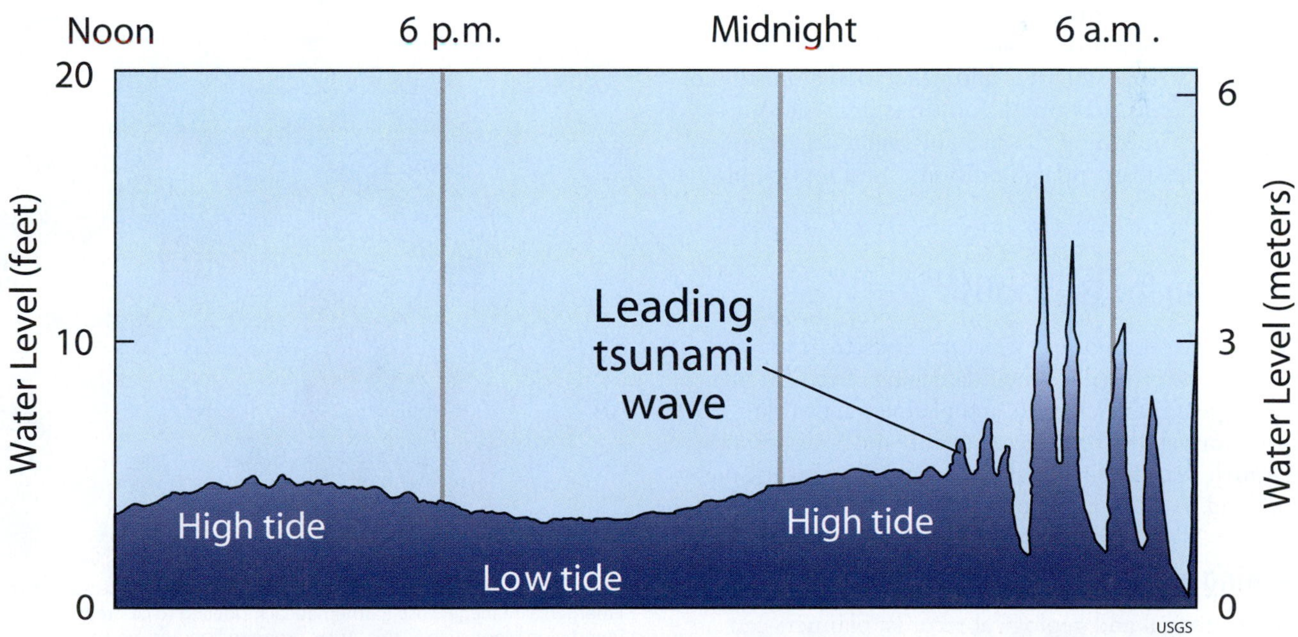

The May 23-24, 1960 tide gauge record for Onagawa, Japan shows a series of eight tsunamis over a 5-hour period. The first waves are not the highest. Note the deep trough preceding the first large tsunami.

USGS

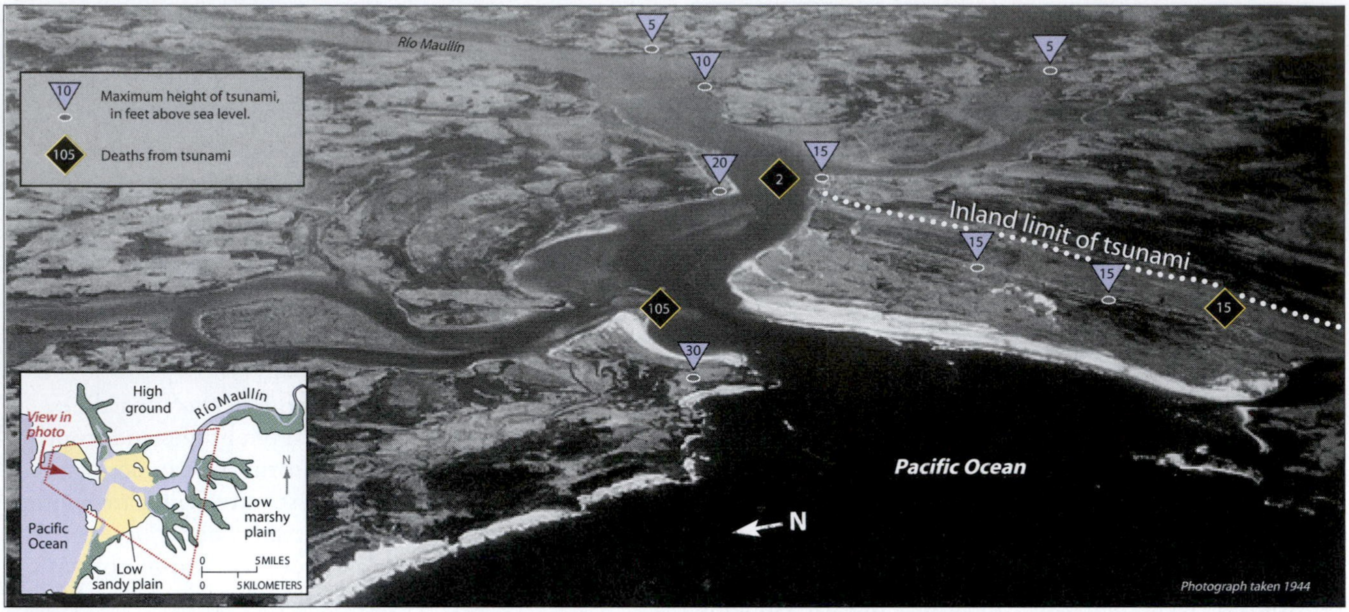

Markings on this 1944 photograph of the area surrounding the mouth of Río Maullín on the Chilean coast show the extent of run-in from the 1960 tsunami. The triangle symbols are labeled with the maximum run-up, in feet, and the diamond symbols indicate where fatalities occurred.

high tsunami with a wavelength of 300 m would hit a 30-m length of seawall with a 135-million-kg (298-million-lb) wedge of water—enough to do incredible damage. Debris caught in the backwash of the leading wave makes the secondary waves even more destructive as they come ashore.

Flooding—In addition to the destructive force of the tsunamis, flooding can kill people, damage property, and spread pollution over a large area.

As shown in the photograph above, run-in from the 1960 Chilean tsunami extended about 3.2 km (2 mi) inland over flat coastal areas and penetrated almost 8 km (5 mi) up the river channels. Run-in is influenced by many factors including the run-up height, orientation relative to the coast, tidal conditions, local topography, and vegetation.

Preparing for tsunamis

Although we cannot prevent natural disasters from occurring, we can reduce their damaging effects through effective planning. Scientists and planners are working to improve our ability to detect tsunamis and issue accurate and timely warnings, to respond appropriately when they occur, and to avoid hazardous situations wherever possible.

Planning ahead

Using historical and geological records, planners can predict where and how often tsunami trigger events are likely to occur. Simulating these events using mathematical models allows them to "see" the effects of tsunamis before they occur and to take corrective measures. Many coastal communities are developing tsunami preparedness plans using inundation models like the one shown below. Public education programs are a big part of these plans. For example, some communities

Oregon Department of Geology/NOAA/PMEL

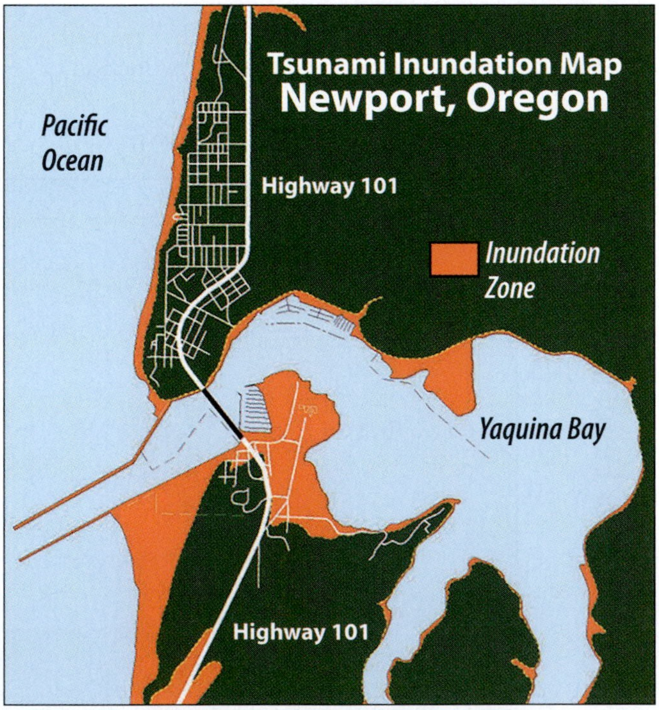

Emergency managers use mathematical modeling to create tsunami inundation maps such as this. Areas prone to flooding are shown in orange, and roads and highways usable as evacuation routes are in white.

now use special signs to alert the public to tsunami hazard zones and evacuation routes.

When high-risk areas are identified before they are developed, appropriate zoning laws ensure that residential developments and large construction projects such as power stations are restricted to higher ground. In areas that have already been developed, retaining walls can be built to provide a higher level of protection. Following the 1993 tsunami, Japan's Okushiri Island built a 15-m reinforced concrete wall to protect vulnerable areas.

Monitoring and warning

Today, Pacific Rim countries use a combination of technology and international cooperation to detect tsunamis. The center of operations for this system is the Pacific Tsunami Warning Center located in Ewa Beach, Hawaii. The Center's objectives are to detect and pinpoint major earthquakes in the Pacific region, determine whether they have generated a tsunami, and provide timely warnings to people living in the Pacific region.

NOAA/PMEL

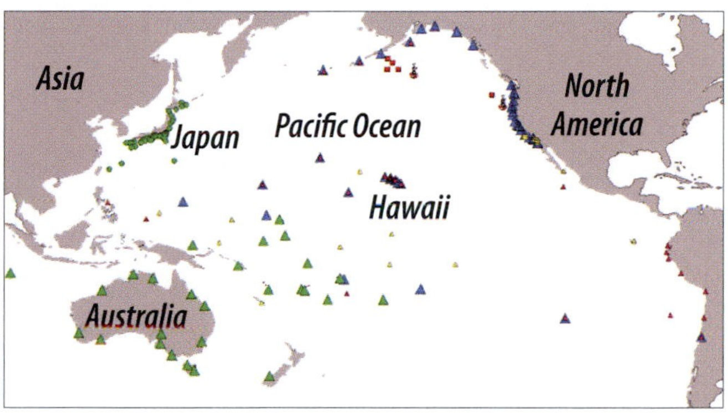

The Pacific tsunami detection system uses 24 seismic stations, 53 tide stations, 52 dissemination points, and 6 DART systems scattered throughout the Pacific Basin.

Deep-ocean tsunamis are difficult to detect. They average less than half a meter high, have wavelengths of hundreds of kilometers, and move at hundreds of kilometers per hour. In 1995, the Pacific Marine Environment Lab (PMEL) developed a system called DART, short for Deep-ocean Assessment and Reporting of Tsunamis. This system uses sensitive detectors to measure water pressure changes from passing tsunamis. It is capable of detecting deep-ocean tsunamis with amplitudes as small as 1 cm.

Large tsunamis are rare, and developing an accurate warning system is a challenging goal. Historically, nearly 75 percent of tsunami warnings have been false alarms. For this reason, people are often hesitant to evacuate their homes and businesses, and their response to warnings in general is poor. Emergency managers in

NOAA/PMEL

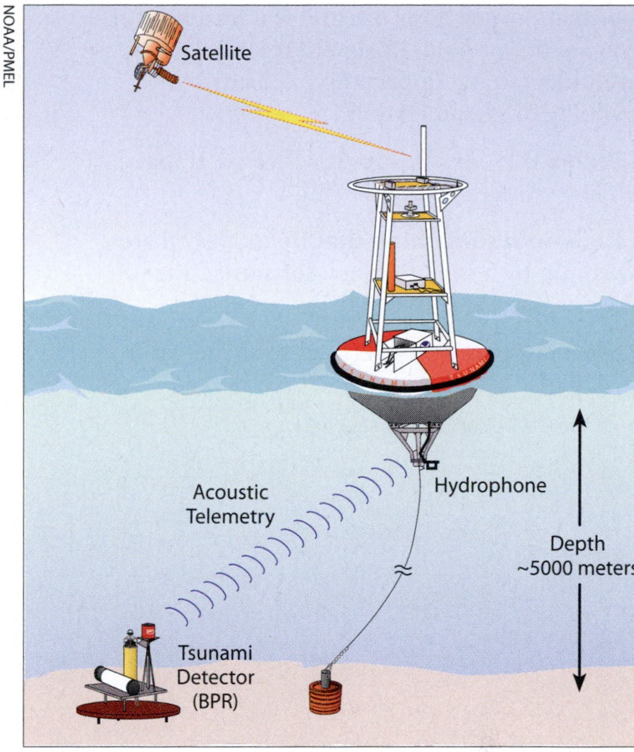

Diagram of a DART station. Placed on the seafloor, a Bottom Pressure Recorder (BPR) transmits pressure information to a surface buoy. The buoy then sends the data to the warning center through a satellite link.

tsunami-prone areas must work constantly to increase public awareness and acceptance of the risks of tsunamis and the emergency plans that are in place.

Surviving a tsunami

By interviewing tsunami survivors, planners have put together these survival tips:

- **Heed natural warnings.** An earthquake is a warning that a tsunami may be coming, as is a rapid fall or rise of sea level.
- **Heed official warnings.** Play it safe, even if there have been false alarms in the past or you think the danger has passed.
- **Expect many waves.** The next wave may be bigger, and the tsunami may last for hours.
- **Head for high ground and stay there.** Move uphill or inland, away from the coast.
- **Abandon your belongings.** Save your life, not your possessions.
- **Don't count on the roads.** When fleeing a tsunami, roads may be jammed, blocked, or damaged.

- **Go to an upper floor or roof of a building.** If you are trapped and unable to reach high ground, go to an upper story of a sturdy building or get on its roof.
- **Climb a tree.** As a last resort, if you're trapped on low ground, climb a strong tree.
- **Climb onto something that floats.** If you are swept up by a tsunami, look for something to use as a raft.
- **Expect company.** Be prepared to shelter your neighbors.

Excerpted from U.S. Geological Survey Circular 1187.

Questions

1. What country is most at risk from tsunamis? Why is this?

2. Why doesn't the East Coast of the U.S. experience tsunamis more often? Are tsunamis possible there?

3. Over geological time, what has caused the most destructive tsunamis in the Hawaiian Islands?

4. How often do devastating tsunamis with run-ups of 15 m or more occur in the Pacific Basin?

5. Why is the first wave of a tsunami often not the most dangerous?

6. Why are tsunamis difficult to detect in the open ocean?

7. How can a community prepare for a tsunami?

8. When is it safe to return to coastal areas after a tsunami?

Investigation 5.4

Tsunami warning

The 1964 Alaska tsunami

Launch ArcMap, and locate and open the **etde_unit_5.mxd** file.

If you encounter unfamiliar GIS terminology or are unable to perform a task, refer to the Quick Reference Sheet in the introduction to this module.

In the Table of Contents, right-click the **1964 Alaska Tsunami** data frame and choose Activate.

Expand the **1964 Alaska Tsunami** data frame.

The 1964 Alaska earthquake was the second strongest earthquake in recorded history and the strongest to strike U.S. territory. The earthquake, shown with the red star symbol, caused extensive damage to Anchorage both through shaking and *liquefaction* (i.e., making liquid) of the soil. The earthquake also triggered a major tsunami that seriously impacted many coastal communities, causing fatalities as far away as Eureka, California.

Turn on the **Plate Boundaries** layer.

Although the epicenter of the earthquake (marked by the red star on your map) was inland, the greatest motion of the seafloor took place some distance offshore, along the Aleutian Trench. An 800-km-long slab of the North American plate was thrust suddenly upward by as much as 40 m as the Pacific plate plunged beneath it.

Turn on the **Tsunami Source** layer. This shows the approximate extent of the displaced seafloor.

Turn on the **Run-up Locations** layer. This layer shows sites that recorded one or more run-up measurements from the tsunami.

To view an animation of the effects of the 1964 Alaska tsunami:

Select the **Trigger Event** layer.

Click on the trigger-event symbol (red star) using the Hyperlink tool. Be patient while the movie loads.

Use the Pause button to examine the movie at various times.

This movie is different from others you have seen. Rather than show a simulated wave, this movie uses colors to show changes from average sea level. Light blues, oranges, and reds are higher than normal sea level; darker blues are lower than normal.

The digital clock at the bottom of the movie screen shows the number of hours and minutes that have passed since the trigger event. The movie begins 27 minutes before the earthquake and ends almost 15 hours later.

Use Pause and Play buttons to return to the frame where the clock reads 00:03. Notice the long red-orange "hill" of water that forms off the Alaska coast. As the movie advances forward in time, the hill spreads out. Follow the light blue leading edge of the tsunami southward (shown as a white dashed line in the figure at left).

Tracking the leading wave

At 1:09 after the earthquake event, the leading wave of a tsunami appears as a light blue arc traveling southward from the source. (The dashed white line does not appear in the movie.)

NOAA/PMEL

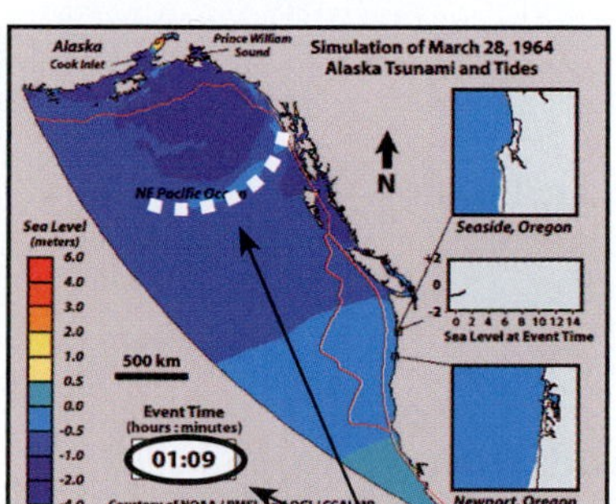

Leading wave at 1:09

Tide gauge plots

Tide gauge plots show the local water level at a gauge as it changes over time. A typical plot, in the absence of factors such as tsunamis, shows a gentle pattern of high and low tides that repeats with a period of around 13 hours.

This plot shows what the Seaside, Oregon tide gauge might have registered if the 1964 Alaska tsunami had never happened. The starting point and time scale of the plot are the same as the graph in the movie.

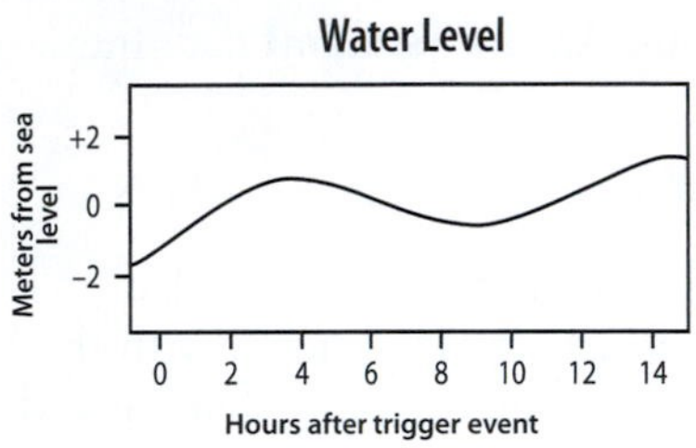

Tides and tsunamis

Tide gauges are devices used to record changes in sea level at coastal locations. Run the movie through several more times, each time focusing your attention on the tide gauge record at Seaside, Oregon. Normal tides would create a regular, gently changing record of the water level like the one shown at left. Watch the leading wave and the tide gauge; you should be able to see the "arrival signature" of the leading wave on the tide gauge readout.

The arrival signature is found where the gauge readout in the movie first differs from the plot in the sidebar to the left.

1. How long did it take the leading wave to reach Seaside, Oregon?

Continue the movie and watch the tide gauge. Earlier you read that tsunamis slosh around an ocean basin (like water in a bathtub). If the waves have enough energy, they can reflect off coastlines and return at a later time. When you reach the end of the movie, answer the following questions.

2. According to the tide gauge, how many secondary waves struck Seaside after the leading wave? (One wave may be very difficult to see in the plot.)

3. Was the leading wave the highest? If not, how many hours after the trigger event did the highest sea-level run-up occur? There were actually two peak waves at Seaside; record them both.

4. Based on what you've seen, is it safe to return to low-lying coastal areas immediately after the leading wave? Explain your answer.

Close the Media Viewer window when you are finished.

Tsunami trigger events

Click the QuickLoad button.

Select **Data Frames**, choose **Tsunami Hazards** from the list, and click **OK**.

Now that you have an idea what tsunamis are and how they travel, you will look at the geologic trigger events that cause them. Any event that displaces a large volume of seawater can generate a tsunami. These trigger events include volcanic eruptions, underwater and coastal landslides, earthquakes, and even (rarely, thank goodness) asteroid impacts. Each of the red star symbols in the **Tsunami Sources** layer represents an event that triggered a tsunami in the 20th century. Next, you will examine these data to find out how common each type of trigger event is.

How often do tsunamis occur?

- Select the **Tsunami Sources** layer.
- Click the Open Attribute Table button.
- Read the total number of events recorded since 1900 at the bottom of the table. (Your answer will be different than the example shown below.)

5. What is the average number of tsunami events recorded each year? (Divide the total number of events by 100 years.)

What causes most tsunamis?

- Scroll across the table to the **Event Type** field. Scroll down the table to see the different types of trigger events.

6. What is the most common type of tsunami trigger event?

- Close the attribute table.

Tsunami warnings

Not all earthquakes produce tsunamis. It's a good idea to warn people of an approaching tsunami, but evacuations are expensive and carry their own risks (panic, looting, etc.). Is there some minimum earthquake magnitude associated with tsunamis? To find out, you will examine the **Tsunami Sources** layer for all events that have an earthquake magnitude greater than zero.

- Click the Select By Attributes button.
- To display the locations of events with magnitudes greater than zero, query the **Tsunami Sources** layer for **("Magnitude" > 0)** as shown in steps 1-6 on the following page. Your query will actually read:

("MAG" > 0)

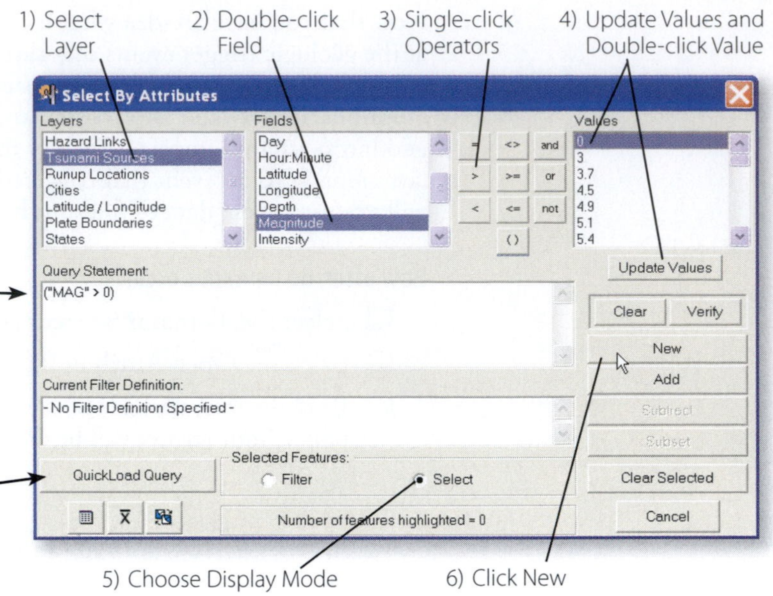

Read query statement here as you enter it.

QuickLoad Query

- Click the QuickLoad Query button and select the **Tsunami EQ Trigger Magnitude** query.
- Click **OK**.
- Click **New**.

If you have difficulty entering the query statement correctly, refer to the **Quickload Query** described at left.

The tsunamis generated by earthquakes with known magnitudes are now highlighted on your map.

To find magnitude statistics for these tsunamis:

Click the Statistics button [X̄] in the Select By Attributes window.

In the Statistics window, calculate statistics for **only selected features** of the **Tsunami Sources** layer, using the **Magnitude** field.

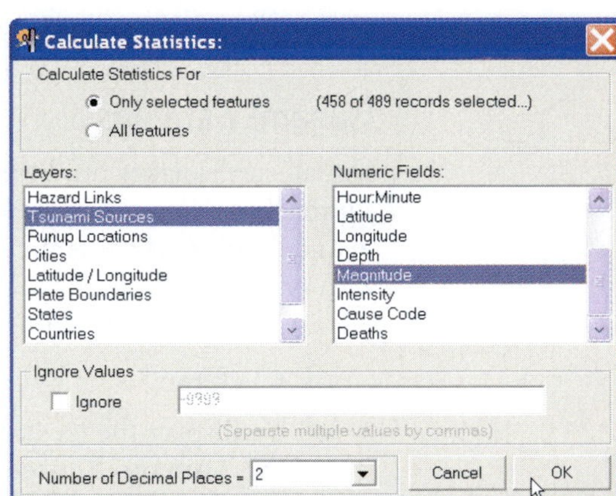

Click **OK**. Be patient while the statistics are calculated.

7. Record the following statistics about the earthquakes that cause tsunamis.
    a. Average magnitude (**Mean**) =

    b. Highest magnitude (**Maximum**) =

    c. Lowest magnitude (**Minimum**) =

Close the Statistics and Select By Attributes windows.

Click the Clear Selected Features button.

Use these statistics to help answer the following questions.

8. Create a list of criteria that you would use to decide whether and when to issue a tsunami warning. Explain each of your criteria.
    a. What size trigger event requires a warning? How close or how far away would it have to occur?

    b. How would your local geography figure into your decision?

    c. When would you issue the warning?

    d. Which officials would you notify? How would you notify them?

    e. What would you tell them?

    f. When would you issue an "all clear" signal?

9. Since 1948, more than 75 percent of tsunami warnings have been false alarms, because it is difficult to predict the impact of a tsunami. Currently a warning is issued each time there is an earthquake of magnitude 6.7 or greater near a coastline or in the open ocean. Do you think it is better to "assume the worst" and send out too many warnings or "assume the best" and send out too few? Explain.

Quit ArcMap and do not save changes.

Investigation 5.5

2004 Indonesian tsunami

On December 26, 2004, at 7:58 a.m. local time, the fourth strongest quake in recorded history (magnitude 9.0) struck off the west coast of Sumatra, part of the island country of Indonesia. The earthquake occurred at the boundary where the India plate subducts beneath the Burma plate. The earthquake caused approximately 1200 km of the plate boundary to uplift several meters. This sudden upward movement of the seafloor displaced trillions of tons of ocean water, generating a tsunami that devastated countries bordering the Indian Ocean and eventually affected all of the world's oceans.

- Launch ArcMap, and locate and open the **etde_unit_5.mxd** file.

If you encounter unfamiliar GIS terminology or are unable to perform a task, refer to the Quick Reference Sheet in the introduction to this module.

- In the Table of Contents, right-click the **2004 Indonesian Tsunami** data frame and choose Activate.
- Expand the **2004 Indonesian Tsunami** data frame.

This data frame shows the location of the main earthquake (the red star), and the locations in which tsunami run-ups were reported.

- Click the Identify tool.
- In the Identify Results window, select the **Affected Countries** layer from the list of layers.
- Use the Identify tool to click on each of the countries listed in the table below in order to find the number of deaths and per capita income for each country.

Per capita [Latin] — by or for each person.

1. Complete the table below using the information obtained from the Identify Results window for each country.

Country	Deaths	Per capita income
Indonesia		
Sri Lanka		
India		
Thailand		
Bangladesh		
Malaysia		
Myanmar		
Somalia		
Tanzania		
Kenya		

- Close the Identify Results window when you are finished.

2. How would per capita income affect a country's ability to respond to a natural disaster of this magnitude?

- Select the **Trigger Event** layer.
- Using the Hyperlink tool, click on the tsunami source off the coast of Indonesia (the red star). Be patient while the movie loads.

The movie shows the initial earthquake and the tsunami as it spread across the Indian Ocean.

3. Did the waves appear to spread evenly in all directions, like circular ripples in a pond when a rock is thrown? Or did they appear to spread more forcefully or vigorously in some directions than in others?

4. If the tsunami does not appear to have spread evenly, how might that have affected the coastlines where the tsunami came ashore? Cite examples from the table.

- Close the Media Viewer window.
- Select the **Run-up Locations** layer.
- Click the Full Extent button to see the entire map.

5. *Teletsunamis* are tsunamis whose waves propagate around the world. Was the 2004 Indonesian tsunami a teletsunami? Why or why not? Use the **Affected Countries** layer as a guide to examine where the tsunami traveled and which coastlines it impacted.

- Quit ArcMap and do not save changes.